I0787958

Processes at the Semiconductor Solution Interface 5

Editors:

C. O'Dwyer
University College Cork
Cork, Ireland

D. N. Buckley
University of Limerick
Limerick, Ireland

A. Etcheberry
L'Institut Lavoisier de Versailles
Versailles, France

A. Hillier
Iowa State University
Ames, Iowa, USA

M. Salazar Villalpando
National Energy Technology Laboratory
Morgantown, West Virginia, USA

Sponsoring Divisions:

 Electronics and Photonics

 Energy Technology

 Physical and Analytical Electrochemistry

Published by
The Electrochemical Society
65 South Main Street, Building D
Pennington, NJ 08534-2839, USA
tel 609 737 1902
fax 609 737 2743
www.electrochem.org

ECStransactions ™

Vol. 53, No. 6

Copyright 2013 by The Electrochemical Society.
All rights reserved.

This book has been registered with Copyright Clearance Center.
For further information, please contact the Copyright Clearance Center,
Salem, Massachusetts.

Published by:

The Electrochemical Society
65 South Main Street
Pennington, New Jersey 08534-2839, USA

Telephone 609.737.1902
Fax 609.737.2743
e-mail: ecs@electrochem.org
Web: www.electrochem.org

ISSN 1938-6737 (online)
ISSN 1938-5862 (print)
ISSN 2151-2051 (cd-rom)

ISBN 978-1-62332-028-7 (Softcover)
ISBN 978-1-60768-379-9 (PDF)

Printed in the United States of America.

PREFACE

The Processes at the Semiconductor Solution Interface 5 (PSSI 5) symposium was held at the 223rd Meeting of The Electrochemical Society from May 12-16, 2013 in Toronto, Canada. The symposium was sponsored by the Electronics and Photonics Division of The Electrochemical Society.

The symposium and the papers in this issue of *ECS Transactions* address topics at the forefront of semiconductor electrochemistry and solution-based processing including electrochemical and semiconductor devices, metallization, porosity formation, electrochemical film growth, and related processes.

This PSSI 5 issue of *ECS Transactions* also contains the 2013 ECS European Section Alessandro Volta Award address by Prof. Jean-Noël Chazalviel.

The editors gratefully acknowledge the authors for their efforts to submit the manuscripts on time, enabling this issue of *ECS Transactions* to be published before the meeting. We thank the organizers, the speakers, especially the awardees and invited speakers, and the session chairpersons for their contributions to the success of the symposium. Finally, we wish to express our appreciation to the staff of The Electrochemical Society for their efforts, which enabled the publication of this issue on a very tight schedule.

C. O'Dwyer, D. N. Buckley, A. Etcheberry,
A. Hillier, M. Salazar Villalpando (Editors)

ECS Transactions, Volume 53, Issue 6
Processes at the Semiconductor Solution Interface 5

Table of Contents

Preface   *iii*

Chapter 1
Europe Section Allesandro Volta Medal Address

(Allesandro Volta Award Presentation) Facts and Challenges in the Electrochemistry   3
and Wet Surface Chemistry of Silicon
  J. N. Chazalviel

Chapter 2
Electrochemical and Semiconductor Devices

(Invited) Semiconductor Nanostructures for Antireflection Coatings, Transparent   25
Contacts, Junctionless Thermoelectrics and Li-Ion Batteries
  C. Glynn, M. Osiak, W. McSweeney, O. Lotty, K. Jones, H. Geaney, E. Quiroga-
  González, J. D. Holmes, C. O'Dwyer

Characteristics Of SnSbSe (SSS) Thin Films Grown by Atomic Layer Deposition for   45
High Performance Phase Change Random Access Memory (PCRAM)
  K. Lee, S. Kang, J. Ku, K. Hong, S. Park

Rechargeable Li-Ion Battery Anode of Indium Oxide with Visible to Infra-Red   53
Transparency
  M. Osiak, W. Khunsin, E. Armstrong, T. Kennedy, C. M. Sotomayor Torres,
  K. M. Ryan, C. O'Dwyer

Chapter 3
Porous Semiconductors and Semiconducting Oxides

(Invited) Cessation of Porous Layer Growth in n-InP Anodised in KOH   65
  R. P. Lynch, N. Quill, C. O'Dwyer, M. Dornhege, H. H. Rotermund, D. N. Buckley

v

TiO$_2$ Nanotubes Formed in Aqueous Media: Relationship between Morphology, Electrochemical Properties and the Photoelectrochemical Performance for Water Oxidation 81
P. Acevedo-Peña, I. González

Chapter 4
Electrodeposition and Semiconductor Metallization

Evidence of Phosphazene Steps Formation on InP by Cyclic Voltammetry Studies and XPS Analyses in Liquid Ammonia (-55°C) 93
C. Njel, A. M. Gonçalves, D. Aureau, D. Mercier, A. Etcheberry

Electroless Metallization of Silicon Using Metal Nanoparticles as Catalysts and Binding-Points 99
S. Yae, M. Enomoto, H. Atsushiba, A. Hasegawa, C. Okayama, N. Fukumuro, S. Sakamoto, H. Matsuda

Changes in the Electrochemical Behavior of Silicon after Platinum Deposition and Ionic Bombardment 105
A. Hervier, D. Aureau, A. Etcheberry

Preliminary Investigations of Ta Surface Chemistry in Aqueous Solutions of TeO$_2$, and the Possible Formation of TaTe$_2$ 113
C. F. Tsang, Y. G. Kim, D. Gebregziabiher, J. L. Stickney

The Nanoporous Metallization of Polymer Membranes through Photocatalytically Initiated Electroless Deposition 123
M. A. Bromley, C. Boxall

Author Index 133

Facts about ECS

The Electrochemical Society (ECS) is an international, nonprofit, scientific, educational organization founded for the advancement of the theory and practice of electrochemistry, electronics, and allied subjects. The Society was founded in Philadelphia in 1902 and incorporated in 1930. There are currently over 7,000 scientists and engineers from more than 70 countries who hold individual membership; the Society is also supported by more than 100 corporations through Corporate Memberships.

The technical activities of the Society are carried on by Divisions. Sections of the Society have been organized in a number of cities and regions. Major international meetings of the Society are held in the spring and fall of each year. At these meetings, the Divisions and Groups hold general sessions and sponsor symposia on specialized subjects.

The Society has an active publication program that includes the following:

Journal of The Electrochemical Society — (JES) is the leader in the field of electrochemical science and technology. This peer-reviewed journal publishes an average of 550 pages of 85 articles each month. Articles are published online as soon as possible after undergoing the peer-review process. The online version is considered the final version and is fully citable with articles assigned specific page numbers within specific issues. The date of online publication is the official publication date of record.

Journal of Solid State Science and Technology — (JSS) is one of the newest peer-reviewed journals from ECS launched in 2012. JSS covers fundamental and applied areas of solid state science and technology including experimental and theoretical aspects of the chemistry and physics of materials and devices. Articles are published online as soon as possible after undergoing the peer-review process. The online version is considered the final version and is fully citable with articles assigned specific page numbers within specific issues. The date of online publication is the official publication date of record.

Electrochemistry Letters — (EEL) is one of the newest journals from ECS launched in 2012. It is dedicated to the rapid dissemination of peer-reviewed and concise research reports in fundamental and applied areas of electrochemical science and technology. Articles are published online as soon as possible after undergoing the peer-review process. The online version is considered the final version and is fully citable with articles assigned specific page numbers within specific issues. The date of online publication is the official publication date of record.

Solid State Letters — *(SSL)* is one of the newest journals from ECS launched in 2012. It is dedicated to the rapid dissemination of peer-reviewed and concise research reports in fundamental and applied areas of solid state science and technology. Articles are published online as soon as possible after undergoing the peer-review process. The online version is considered the final version and is fully citable with articles assigned specific page numbers within specific issues. The date of online publication is the official publication date of record.

Electrochemical and Solid-State Letters — (ESL) was the first rapid-publication electronic journal dedicated to covering the leading edge of research and development in the field of solid-state and electrochemical science and technology. ESL was a joint publication of ECS and IEEE Electron Devices Society. Volume 1 began July 1998 and contained six issues, thereafter new volumes began with the January issue and contained 12 issues. The final issue of ESL was Volume 16, Number 6, 2012. Preserved as an archive, ESL has since been replaced by SSL and EEL.

Interface— *Interface* is an authoritative yet accessible publication for those in the field of solid-state and electrochemical science and technology. Published quarterly, this four-color magazine contains technical articles about the latest developments in the field, and presents news and information about and for members of ECS.

ECS Meeting Abstracts— *ECS Meeting Abstracts* contain extended abstracts of the technical papers presented at the ECS biannual meetings and ECS-sponsored meetings. This publication offers a first look into the current research in the field. ECS Meeting Abstracts are freely available to all visitors to the ECS Digital Library.

ECS Transactions— (ECST) is the online database containing full-text content of proceedings from ECS meetings and ECS-sponsored meetings. ECST is a high-quality venue for authors and an excellent resource for researchers. The papers appearing in ECST are reviewed to ensure that submissions meet generally-accepted scientific standards. Each meeting is represented by a volume and each symposium by an issue.

Monograph Volumes — The Society sponsors the publication of hardbound monograph volumes, which provide authoritative accounts of specific topics in electrochemistry, solid-state science, and related disciplines.

For more information on these and other Society activities, visit the ECS website:

www.electrochem.org

CHAPTER 1

EUROPE SECTION ALLESANDRO VOLTA MEDAL ADDRESS

2

Facts and Challenges in the Electrochemistry
and Wet Surface Chemistry of Silicon

J.-N. Chazalviel[a,*]

[a] Physique de la Matière Condensée, CNRS, Ecole Polytechnique,
91128 Palaiseau, France

Silica being insoluble in non-fluoride medium, anodic oxide films
are readily formed on silicon. Dissolution of any oxide film in HF
leaves a surface covered with hydrogen. Anodization in HF leads
to the formation of porous silicon, the hydrogen coating staying
present during the process. At higher potentials, electropolishing is
observed, with the presence of a thin oxide layer on the surface. At
still higher potentials, mesoporous oxide films are obtained. In
non-aqueous media, trace amounts of water unavoidably lead to
irreversible formation of a silica layer. In methanol a more stable
behavior is obtained, due to chemical surface modification by
grafting of methoxy groups. Other modifications of the silicon
surface include the formation of covalent Si-C bonds, which holds
promises for applications, especially sensors. In the near future,
challenges in silicon electrochemistry include a better
understanding of several fundamental issues, and also improved
preservation of the surface against oxidation.

Introduction

The second half of the twentieth century has seen the advent of silicon as the key material
of microelectronics. Though the most critical processing steps in established silicon
technology are performed in dry conditions, this has stimulated interest in the
electrochemistry and wet surface chemistry of that material. After the early studies in the
fifties (1-4), this interest has been revived by several landmark events. In the mid-
seventies, a surge of interest for semiconductor electrochemistry was triggered by the
hope to realize the photoelectrochemical conversion of solar energy with semiconducting
photoelectrodes (5). Though silicon does not appear as an ideal material in this context,
fundamental knowledge on the silicon/electrolyte interface has greatly benefited from this
effort (6). In the early nineties, a revival of interest was triggered by porous silicon, a
long-known material obtained by electrochemical anodization of a single-crystal silicon
surface in HF (2,7,8). The astonishing finding that porous silicon may be strongly
photoluminescent in the visible range aroused an unprecedented effort in the
electrochemistry of silicon in fluoride medium (9-11). Finally, in the late nineties, a new
interest arose on the surface chemistry of silicon, on one hand because it can be
controlled down to atomic level, providing an ideal benchmark for surface chemistry
studies, and on the other hand because the subject is very promising in terms of
applications, especially in the field of sensors (12-15).

---

* Now on retirement. Present address: 25 rue Eugène Combes, 19800 Corrèze, France.

In this paper, we will attempt to briefly review the main knowledge on the electrochemistry and surface chemistry of silicon, with an effort to distinguish what seems firmly established from the more controversial points. In a last part, we will try to identify some of the challenges of the domain for the near future. Of course, such a brief review cannot be exhaustive. We will attempt to give a fair summary of the present state of knowledge, but this review will still largely reflect our own interests.

Facts in Silicon Electrochemistry

Silicon is a semiconductor; hence, in any electrolyte, silicon electrodes usually exhibit rectifying behavior. Namely, in the dark, moderately doped n-Si electrodes lead to a very small anodic current and p-Si electrodes lead to a very small cathodic current (6). This behavior disappears whenever minority carriers are created near the surface, either by bandgap illumination or by junction breakdown under strong reverse bias, a process which takes place at increasingly small potentials as doping level is larger (16). While keeping in mind these classical aspects, a more specific feature of silicon, related to its chemistry, appears most important, namely the insoluble character of silica, except in fluoride medium. This leads us to distinguishing three major kinds of electrolytes: aqueous non-fluoride media, fluoride media, and non-aqueous media.

Silicon in aqueous non-fluoride electrolytes

Silica is a good insulator, and the surface of a silicon electrode can be easily passivated with a thin layer of silica. Early studies of electrochemical oxide growth on silicon electrodes under galvanostatic control indicated an oxide thickness of 0.4 nm/V (17). Silica layers up to ~100 nm thickness could be obtained in this way. However, these anodic oxides tend to incorporate ions from the electrolyte (18). Hence, they exhibit poor electronic properties: high concentration of interface states and fixed charges. This makes them unsuitable as gate insulators in MOSFET technology. Interestingly, it has been shown that anodization of silicon in ultrapure water leads to a silicon/oxide interface of much higher electronic quality (19). However, this method has not as yet emerged as a practical route to gate oxide fabrication.

The first stages of anodic oxidation of an n-Si electrode have been followed by transient studies monitoring the small dark current when the electrode is put into contact with the electrolyte. Systematic determinations of the flatband potential by impedance measurements and Mott-Schottly analysis have also been made (20). These studies indicate that the small dark current flows through the intermediate of interface states, which appear associated with the oxidation process, and electron transfer from the interface states to the conduction band takes place by activated transfer over the Schottky barrier. The flatband potential appears to be pH dependent, with a slope of ~30-40 mV per pH unit (20,21), and the Schottky-barrier height follows the same trend, with a lower slope of ~10 mV per pH unit (20). Direct evidence for the presence of the interface states is obtained from a peak in the capacitance/potential curves and from the presence of a small subbandgap photocurrent, associated with photoassisted electron transfer from the occupied interface states to the conduction band (22).

The cathodic behavior of silicon electrodes has been the subject of some studies, especially in view of the photoelectrochemical generation of hydrogen (23-25). p-Si appears as a poor candidate for that purpose. The poor catalytic properties of the silicon surface can be remedied by the deposition of transition-metal islets (23,24), but the gain in phopotential is low, and oxidation of the surface in water is still a problem for long-term performance. Furthermore, hydrogen can be incorporated into the silicon lattice, leading to a new kind of interface states prejudicial to the electrical quality of the junction (26-28). Note that the same phenomenon may take place at zero current during chemomechanical polishing (29).

Silicon is efficiently dissolved at open-circuit potential (OCP) in strongly basic electrolytes, a chemical reaction accompanied with hydrogen evolution. The dissolution rate of silica remains much smaller in the same media. Interestingly, the dissolution mechanism of silicon appears highly anisotropic (much slower etching of the {111} planes), and is inhibited at high dopings. These features have been widely exploited for silicon micromachining (30). The etching mechanism has been discussed in light of various studies (effect of an applied potential (31), in situ ellipsometry (32) and scanning tunneling microscopy (33)). For applied potentials slightly negative of OCP, etching of the {111} terraces can be fully inhibited, leading to dissolution of {111} planes by a step-flow mechanism. For increasingly positive potentials, the etching becomes more isotropic, till oxidation takes place, leading to passivation (Flade potential). The change in behavior around OCP has been interpreted in terms of the coexistence of two mechanisms: a purely chemical mechanism, strongly anisotropic, and a zero-current electrochemical mechanism, exhibiting less anisotropy than the former one (34). The transition between the active dissolution state and the passive state leads to the appearance of electrochemical transients that may still deserve investigation (35).

Silicon in fluoride electrolytes

By contrast with alkaline non-fluoride media, hydrofluoric acid dissolves silica very rapidly, but silicon is essentially not attacked. An oxidation followed by an HF rinse then represents a simple way to prepare clean, oxide-free silicon surfaces.

Silicon at OCP in fluoride medium. A most remarkable discovery is that HF-rinsed silicon surfaces are virgin of any oxide and remain so on a time scale of a few minutes to an hour, even when handled in atmosphere. This surprising passivation against oxidation has been found to be due to the presence of a covalent hydrogen coating on the surface. The presence of hydrogen was first discovered in the seventies by Harrick and Beckmann (36), but this result was largely ignored. It was rediscovered in the eighties (37-39), and a few years more were necessary before general agreement is reached (40). The fact that hydrogen, rather than fluorine, is left on the surface (though the Si-F bond is much stronger than Si-H) can actually be understood as a kinetic effect: the limiting step of (very slow) Si etching in HF is attack of the SiH groups by the fluoride species; hence, the steady-state of the surface in contact with the liquid is essentially the hydrogenated state (41). However, surfaces obtained by rinsing in HF are atomically rough, exhibiting SiH, SiH_2 and SiH_3 species, with a variety of environments (40). For the case of {111} surfaces, it was found that using slightly basic ammonium fluoride instead of HF (a combination of the properties of the alkaline and fluoride media) leads to anisotropic etching with formation of atomically flat hydrogenated terraces, where the hydrogen is exclusively in the form of monohydride; i.e., the dangling bonds of the {111} terraces, all pointing in the direction perpendicular to the surface, are saturated with hydrogen (42,43). These {111} surfaces, unreconstructed (44), perfectly ordered on distances as large as 100-1000 nm, and virgin of any surface states in the bandgap (45), represent an important achievement in terms of surface control down to atomic scale. The treatment has been further optimized (46), leading to a well-defined starting point for many studies (see, e.g., last section).

Silicon dissolution at moderate potential: porous silicon. The voltammetry of p-Si or illuminated n-Si in fluoride electrolyte exhibits several regions in the anodic range (47-51). In a first potential range (A) close to OCP, the current increases with increasing potential, with a Tafel slope of 60 mV/decade (52). The higher the fluoride concentration, the larger the current that may be reached in this region. In this range, the bands at the surface for p-Si are under slight depletion conditions (52), the dissolution is divalent and accompanied with hydrogen evolution $[Si-2e^- \rightarrow Si(II) \rightarrow Si(IV)+H_2]$ (3), the impedance diagrams exhibit a low-frequency inductive loop (53,54), and the surface recombination velocity is low, though it increases with increasing current density, indicating an increasing concentration of surface states within the bandgap (55). For the case of illuminated n-Si, photocurrent multiplication is observed, which gives evidence for the presence of electron-injecting intermediate states (56,57). A related current peak is observed when an oxidized surface is exposed to fluoride electrolyte. This phenomenon indicates electron injection by the intermediate states when oxide dissolution is nearly complete (56,58,59). In-situ infrared spectroscopy of the surface in the fluoride electrolyte indicates that the surface remains essentially hydrogenated in this potential range (60). It is found that silicon dissolution is not complete, but a *porous silicon* layer is left on the surface. Under galvanostatic conditions, the thickness of this layer increases linearly with anodization time (7,8). Thicknesses up to several tens and even hundreds of micrometers can be obtained.

Porous silicon (PS) is actually a new material, which has been the subject of hundreds of studies (7,8,61). In practice, the anodic treatment is performed under galvanostatic conditions in concentrated HF, generally in the presence of ethanol or another surfactant, aimed at avoiding the presence of hydrogen bubbles sticking to the surface. For p-Si, the layer is generally micro- or mesoporous, with a characteristic pore size increasing at high doping concentrations (7). At low doping concentrations, the initially microporous morphology turns to macroporous as the layer is made thicker (Fig. 2) (62). For n-Si, a variety of experimental conditions may be used: anodization in the dark under space-charge breakdown conditions (16), anodization at fixed potential under front (63) or back illumination (64). The morphology depends on the detail of these conditions, though macropores are more often obtained on n-Si than on p-Si (62,65). The PS surface comes out hydrogenated, but it may be oxidized either slowly by ageing or more quickly by various treatments (thermal, anodic, or photochemical oxidation). Chemical modifications have also been explored (66). PS has attracted much interest due to the variety of morphologies that may be obtained. Depending on the conditions, the macropores may grow either parallel to the current lines (actually normal to the envelope of the pore front (67)) or along preferred crystallographic directions (<100> favored), and prepatterning may be used, opening new routes for microfabrication (64,68).

Finally, it has been found that microporous PS is strongly photoluminescent in the visible range, a very surprising fact since Si is an indirect-bandgap semiconductor with a gap in the infrared range (9). After many conflicting studies, it is generally agreed that this result is essentially due to quantum confinement of the electronic states in the silicon nanostructures, though surface effects may also play a role (69). A key experiment in this respect is the comparison of PS made from p-Si with PS made from amorphous hydrogenated silicon (70). In either case a visible PL is observed, but the behavior of this photoluminescence when the nanostructure is dissolved in HF is strikingly different: for PS from p-Si, the luminescence shifts from the near-infrared to the red and the yellow, indicating increased confinement effects when the structure is thinned down before full dissolution, whereas for PS from a-Si:H no shift is observed, due to the fact that the electronic states responsible for the luminescence are already localized by disorder in the initial (amorphous) material (Fig. 1) (70). There has been many attempts to elaborating electroluminescent structures from PS, but the best results in terms of efficiency and stability are still somewhat below expectation (71). The possibility of elaborating silicon nanostructures by chemical etching in the presence of an oxidant, either a redox couple in solution (72,73) or a metal/ion couple (metal-assisted dissolution) (74-76) may open new routes in this respect. Note that the latter method also allows for the elaboration of nanowire arrays, a morphology opening new fields of application.

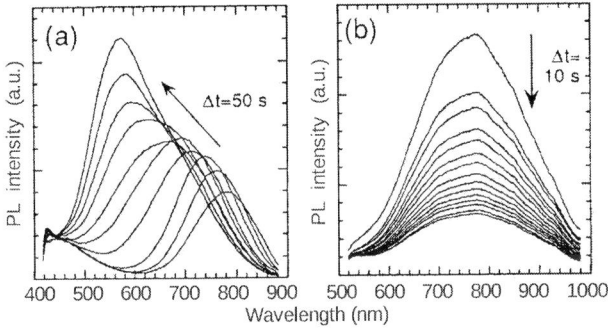

Figure 1. Time evolution of the photoluminescence of microporous silicon immersed in HF. Porous silicon is made from crystalline silicon (a) and from amorphous hydrogenated silicon (b). The etching of the nanostructures results in a blue shift of the band in (a), due to quantum confinement effects. The shift is absent in (b), because the electronic states are already localized by disorder in the initial material (70).

Meanwhile, the very formation mechanism of porous silicon remains a controversial issue. There is no doubt that various mechanisms must be considered depending on the type of porous silicon (from p-Si or n-Si, microporous or macroporous,...). It is generally agreed that macropores formed on n-Si at high potential in the dark are due to the fact that avalanche breakdown of the space-charge is favored at the pore tips, where the electric field is larger (16,77). Also, the role of the space-charge layer thickness in determining the pore spacing, though it may not be an absolute rule (78), is generally recognized (79). However, the origin of micropores is less clear. For p-Si, several physical and chemical mechanisms have been invoked for explaining current enhancement at the pore tips and passivation of the pore walls. These include Schottky-barrier lowering (80,62), increased diffusion current (81), quantum confinement of the electronic states (82), hydrogen passivation (83), defects (84,85), reaction intermediates (86,87)... We have shown that a model involving some of these effects is able to account for the morphologies observed at p-Si, including the transition to macropores for low-doped material (88,89), but there is as yet no general agreement on a single model (90).

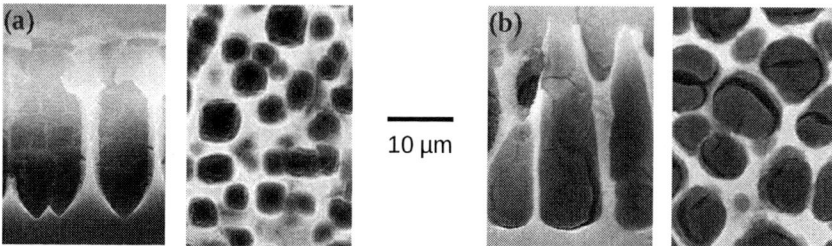

Figure 2. Example of two complex morphologies: porous silicon made from low-doped p-Si {100} (400 Ωcm, 30 mAcm^{-2}). (a) HF 10%: "crystallography-driven" macropores, facetted and aligned along {100}; (b) HF 20%: "current-line driven" macropores, rounded and filled with microporous silicon. Note the top microporous layer, present in both cases (removed by polishing for acquiring the top views). These morphologies can be accounted for quantitatively (88,89).

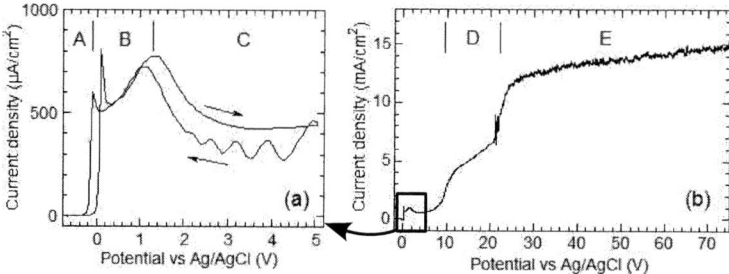

Figure 3. Typical voltammogram of p-Si in fluoride medium, here 1M NH$_4$Cl + 0.025M NH$_4$F + 0.025M HF (fluoride concentration 0.05 M, pH 3). (a) is a zoom of (b).

Silicon dissolution at higher potentials: electropolishing and the oscillations. When potential is increased beyond the regime of porous silicon formation (region A), a small current peak is observed, followed by a first current plateau (region B), a second current maximum, and a second plateau (region C), extending typically through the 1.5-10 V vs SCE range (Fig. 3a). The magnitude of these currents increases with increasing fluoride concentration and decreases when pH increases above 3. The reaction is under mixed control conditions and follows the Koutecky-Levich law $1/J = 1/J_k + 1/J_d$, where J_k is the kinetic contribution and J_d is the diffusion current corresponding to the supply of fluoride species (91-93). The voltammograms are essentially the same for p-Si and for illuminated n-Si, except for a small shift in potential corresponding to the photopotential (49,51). The presence of oxide on the surface can be detected from the first current peak (94). In region (B), often referred to as the "first electropolishing plateau", the dissolution gradually changes from divalent to tetravalent (3), the impedance diagrams exhibit a low-frequency capacitive loop (53,54), the surface appears coated with a "wet-oxide"/hydroxide film (95), the current is slightly orientation dependent (50,96,97), and the surface recombination velocity is high (98). In this region, the surface remains rather rough on the nanometric scale (99). In region C ("second electropolishing plateau"), the bands at the surface turn to accumulation conditions (100), the current decreases then slowly re-increases with increasing potential (47-51), the dissolution is nearly tetravalent (101), there is an oxide layer whose thickness, of a few nm, increases with increasing potential, reaching a value on the order of 10 nm at 5 V (95), and the surface is under good electropolishing conditions (99). Interestingly, the electrochemical current in this region tends to exhibiting an oscillatory behavior, which has stimulated much fundamental interest (54,102-124).

These oscillations are observed either under galvanostatic conditions (potential oscillations) or more often under quasi-potentiostatic conditions (current oscillations in the presence of a small series resistance, which may just consist of the series resistance of the silicon bulk, the electrolyte, and/or the back-contact). Various physical quantities have been found to oscillate at the same time as the current: thickness of the interfacial oxide layer (103,106,112-114,122), optical properties of the oxide (112), surface hole concentration (111,112), interface strain (109,116), surface recombination velocity (110),... These oscillations exhibit a marked difference with most electrochemical oscillations. Namely, under true potentiostatic conditions (that is, if care is taken to avoid any series resistance), a stable current is obtained, as it is common among electrochemical oscillating systems (105). However, unlike the case of most of these systems, here any perturbation of this *stable steady state* (potential step or change in the potential sweep rate, change in electrode rotation rate,...) leads to excitation of a current oscillation, which is slowly damped after the perturbation is stopped (Fig. 4a) (103). This behavior is better termed resonant rather than oscillatory (54). The electrochemical impedance exhibits indeed a resonant behavior at the frequency of the "spontaneous" oscillation and at its overtones (though the measurements are made in the linear-response regime) (Fig. 4b) (54). It indicates that the oscillation is present indeed, "hidden", even in the stable steady state. The key for this original behavior has been inferred from the experiments represented in Fig. 4c-d (103). A strong oscillation is triggered, for example by applying a small potential excitation at the resonance frequency. Then, at some time during the oscillation, the potential is stepped to a value close to OCP, and the current is monitored as a function of time. The recording exhibits a peak, which has been attributed to electron injection from dissolution intermediates at the end of the dissolution of the oxide (56,58,59). The delay at which the current peak occurs can then be taken as a qualitative indication of the oxide thickness at the time when the potential was stepped. The variation of this delay with the stepping time gives a direct evidence for the oscillation of the oxide thickness during an oscillation period. However, if the electrode is brought to the steady state, stepping the potential to the same value leads to the observation of a much broader current peak, which appears indeed as the average of the peaks obtained at various stepping times during an oscillation period (103). This gives direct evidence that the oxide thickness in the steady state is *distributed*, and that this distribution is consistent with a sustained oscillation being present at any point of the surface, the various points of the surface being uncorrelated in phase. Turning this idea of small "self-oscillating domains" into a quantitative form has allowed us to account for the very complex impedance diagrams quantitatively and with a very small number of free parameters (104). There is as yet no direct visualization of these domains. The very origin of the oscillation has been the subject of many proposals, two of which have been put in a quantitative shape. For Lewerenz et al., the oscillation arises from periodic stress-induced breakdown of the oxide layer (107,118,119). For Föll et al, dielectric breakdown is the key mechanism (114,117,121). Either model accounts for many though not all of the experimental observations, leaving in our opinion the issue imperfectly solved.

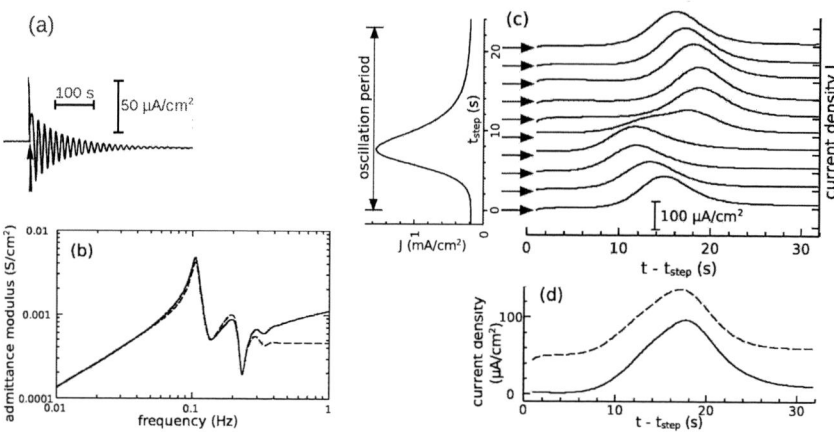

Figure 4. Resonant behavior in region (C). (a) Current transient obtained upon stepping the potential from 3.0 V to 3.1 V. (b) Impedance plot (Bode form, solid line) and the fit (dashed line) obtained from the local oscillator model (54,104). (c) Anodic transient monitoring oxide dissolution at various times during an oscillation period (here forced oscillation at 4.5 V). (d) The solid curve represents the transient obtained for an electrode in the steady state, which coincides with the average of the transients at various times during the oscillation (dashed line, shifted for clarity), a strong evidence for the presence of local oscillators (103).

Silicon dissolution at very high potentials: porous silica. Above a typical potential of 10 V, the current again exhibits a steep increase (Fig. 3b). However, strong oxygen evolution is found to occur only above 20 V (region E). In between (region D), silicon dissolution is speeded up. Combined electrochemical impedance/infrared measurements have shown that this is due to the formation of a rather thick (~ 100 nm) porous oxide layer (125). Still thicker layers (up to micrometer thickness) may be obtained when starting from a dilute fluoride electrolyte with a neutral or slightly basic pH. Silica is almost insoluble in such a medium, but the anodic oxidation of silicon at the pore bottoms turns the medium locally acidic, allowing for partial dissolution of the oxide formed, and leaving a porous oxide behind (126-128). A similar mechanism has been invoked to explain the formation of porous titania (129). For the case of silicon, the mechanism has been put into a quantitative shape, allowing to reproduce the complex shape of the voltammograms, which exhibit bistable behavior (127,128).

Unlike porous alumina (130) and porous titania (129), the obtained silica films are mesoporous, with no well-defined pores but rather a spongy structure. Furthermore, the films formed at higher potentials exhibit a macrostructuration with various morphologies (waves, bowls, labyrinths), which has been attributed to partial dissolution of the film due to acidification caused by the onset of oxygen evolution (131). Finally, careful studies of the mesoporous oxide, using X-ray reflectivity (132) and SEM examination after filling with electrodeposited nickel (133), reveal the presence of a stratified structure on the scale of ~ 10 nm. It has been demonstrated that this periodic structure is associated with the oscillatory phenomenon observed in the 2-10 V potential range, which survives in the regime of porous oxide formation (125,128,133).

Silicon in non-aqueous electrolytes

The addition of non-aqueous solvents to fluoride electrolytes has been used as an extra handle to influence the morphology of porous silicon or porous silica. For example; it has been found that the use of solvents such as dimethylformamide or dimethylsulfoxide allows for elaborating macroporous silicon with pores of a higher aspect ratio and a better defined geometry, the growth along the <100> directions being strongly favored (134,135). This can be attributed to a higher anisotropy of the electropolishing currents $(J\{111\}<<J\{110\}<J\{100\})$ in the presence of these solvents. Also, the use of glycol has been reported to be beneficial for the elaboration of porous silica (136). However, in these studies except a few ones (137), the role of the organic solvent was rather that of an additive, the medium remaining largely aqueous.

In the context of photoelectrochemistry, there has been studies of silicon in anhydrous non-aqueous solvents (138-140). The aim contemplated was to realize a photoelectrochemical cell with a redox system, and avoiding silicon oxidation (passivation) or dissolution. The extreme sensitivity of the silicon surface to trace amounts of water made these attempts largely unsuccessful. Especially, a study of the flatband potential in nominally anhydrous acetonitrile electrolyte in the presence of various redox systems revealed that a freshly prepared hydrogenated silicon surface leads to a Schottky-barrier height depending on the redox system as expected, but ageing on an hour time scale leads to a barrier height essentially independent of the redox system, an indication of strong Fermi-level pinning (140). Complementary studies confirmed that the initial surface is essentially virgin of oxide and electronically clean, but ageing in contact with the electrolyte leads to oxidation with the formation of interface states in the bandgap (140). These interface states were demonstrated to play the role of intermediate in the electrochemical transfer for simple redox systems (141).

However, the case of alcohol solvents and especially methanol appeared singular in this respect. In contrast to the case of acetonitrile and other non-aqueous solvents, n-Si/electrolyte junctions prepared in methanol lead to stable junctions. The Schottky-barrier height and the photopotential depend on the potential of the redox system as expected for an ideal semiconductor/electrolyte junction (142), allowing for the elaboration of stable and efficient n-Si/redox system photoelectrochemical cells (143). It has been shown that this good result is actually due to a chemical modification of the silicon surface by the spontaneous grafting of methoxy groups (144). This provides chemical stabilization of the surface against oxidation, and at the same time the associated dipole leads to a shift of the flatband potential, increasing the photopotential and the performance of the cell (142-144). The spontaneous grafting of methoxy groups is one of the first examples of the beneficial effect of a chemical modification of the silicon surface by organic grafting. This grafting can actually be made much faster under anodic polarization, as it was demonstrated later for porous silicon (145). Since this work was done, organic modifications of the silicon surface, using electrochemical or wet chemical routes, have developed into a wide field, which will be briefly summarized now.

Facts in Modification of the Silicon Surface

The organic modifications of the silicon surface have been reviewed in several recent papers (13-15), and we will mention here only a selected set of results. The field of applications involves the protection of electrodes or the prevention of silicon oxidation, but the major field of applications is that of sensors. The aim is then to immobilize a specific chemical or biochemical species on a surface (probe molecule). The choice of silicon as a substrate is one among others. It is largely motivated by the perfect control of the initial surface down to atomic scale. The first step is then the preparation of a hydrogenated surface. In a second step, the surface hydrogens are substituted by organic species. Anchoring of the organic species is generally realized through an Si-O-C bridge or a direct covalent Si-C bond, the latter method leading to an attachment less prone to hydrolysis, hence more robust in atmospheric or aqueous environment. This substitution reaction can be carried out electrochemically, but also chemically, using thermal, photochemical, or catalytic activation.

Anchoring through an Si-O-C bridge

In the same way as methanol, alcohols can react with the hydrogenated silicon surface, according to the reaction $\equiv SiH + HOR \rightarrow \equiv SiOR + H_2$. The reaction is usually carried out using thermal activation (typically 100°C, 10 h) (146). The alcohol must be anhydrous, otherwise silicon oxidation takes place in parallel. The fraction of substituted hydrogens is limited by steric hindrance among the alkoxy groups. In practice, at an ideal {111}-SiH surface, a substitution of 50% may be expected. The same type of modification may be reached by using aldehydes instead of alcohols: $\equiv SiH + O=CH-R \rightarrow \equiv SiOCH_2R$, which is reported to lead to slightly more compact layers (146).

Anchoring through a direct Si-C bond

Electrochemical methods. Electrochemistry provides a simple method to generate free radicals, which can abstract a hydrogen atom from the hydrogenated silicon surface, making it very reactive for grafting. A popular example of such methods is the grafting of aromatic groups by cathodic (photocathodic on p-Si) reduction of diazonium compounds in aqueous medium: $R\phi N_2^+ + e^- \rightarrow R\phi^\bullet + N_2$ (radical generation), followed by $\equiv SiH + R\phi^\bullet \rightarrow \equiv Si^\bullet + R\phi H$ (hydrogen abstraction), and $\equiv Si^\bullet + R\phi N_2^+ + e^- \rightarrow \equiv Si\phi R + N_2$ (grafting) (147,148). The advantage of these methods is that they are very fast as compared to chemical methods (the grafting itself can be carried out in a few seconds). Their disadvantage is that the free radicals are very reactive, and they may often rearrange or react with species in solution (leading to undesired species) or else react with the already grafted layer; if such is the case, the grafting is not self-limited to a monolayer, and a polymeric layer is obtained instead. For the case of diazonium precursors, and though formation of a polymeric layer is possible, it has been shown that proper control of the Faradaic charge may enable one to stop the modification at the monolayer level (149). Since the reaction is performed at negative potentials, silicon oxidation is not too severe. It may be completely avoided by working in a non-aqueous solvent (149). Finally, one must note that the energetics of the reaction depend on the substituent R. For example, for R = NO_2 (nitrobenzenediazonium), the standard potential of the first step is shifted in the positive direction, so that the reaction can take place spontaneously at OCP, leading to spontaneous grafting at zero current (150).

Another example of electrochemical grafting is the anodic (photoanodic on n-Si) grafting of alkyl chains from Grignard precursors in ether or tetrahydrofuran: $RMgX \rightarrow R^{\bullet} + MgX^{+} + e^{-}$, followed by $\equiv SiH + R^{\bullet} \rightarrow \equiv Si^{\bullet} + RH$ and $\equiv Si^{\bullet} + RMgX \rightarrow \equiv SiR + MgX^{+} + e^{-}$. Since an alkyl radical cannot abstract a hydrogen from an alkyl group, the grafting is self-limited to a monolayer (actually 50% of the hydrogen sites on {111}-SiH, due to steric hindrance effects). Furthermore, the presence of the Grignard insures that the medium is perfectly anhydrous, which prevents any oxidation of the silicon surface during the grafting (151). The case of methyl is special, because the small size of the methyl groups allows for 100% substitution of the hydrogens at the {111}-SiH surface (152). This provides a robust protection of the surface. In the context of the development of high-K oxide technology for next-generation MOSFETs, the unique thermal stability of the methylated Si surface has been proposed as a method to avoid formation of an interfacial SiO_2 layer during MOCVD deposition of HfO_2 on silicon (153). Aromatic or unsaturated groups (vinyl, ethynyl) can also be grafted from Grignard precursors (154,155). However, in that case the aryl, enyl or ynyl radical formed is able to abstract hydrogen from most solvents and to undergo reactions with the already grafted groups. Solvent attack may be avoided by exchanging the ether with 1,2-dichlorobenzene, but anyway the reaction is not self-limited to a monolayer. It has been shown that using 4-chlorophenyl magnesium bromide as the precursor orients the attachment of new phenyl groups in the para-position, providing a simple method to form poly-paraphenylene layers anchored to silicon (154).

Other electrochemical grafting methods include the cathodic grafting of alkyl chains from halogenoalkanes, a mechanism proceeding though the formation of alkyl radicals, in the same way as the anodic decomposition of alkyl Grignards (156), and the anodic and cathodic decomposition of alkynes (157).

Chemical methods. As compared to electrochemical methods, chemical methods are sometimes preferred because they do not require taking a back-contact. For example, grafting of alkyl chains from Grignard precursors may be done thermally (158,159), a reaction which has been demonstrated to be actually a zero-current electrochemical reaction in the presence of alkyl halide in the Grignard (160). Alternately, a two-step procedure (surface halogenation followed by treatment with a Grignard) has been proposed (161). However, the major advantage of chemical routes as compared to electrochemical (free-radical) methods is when the immobilization of fragile chemical functions is desired. The most studied chemical method for the immobilization of organic species on silicon is through the hydrosilylation reaction of double or triple bonds: $\equiv SiH + H_2C=CH-R \rightarrow \equiv Si-CH_2-CH_2-R$ or $\equiv SiH + HC\equiv C-R \rightarrow \equiv Si-CH=CH-R$. Here, R may bear various terminal functions, especially esters or carboxylic acids (alcohols, aldehydes or amines are not advisable, because they may also react with the SiH surface through the functional group). The reaction may be activated thermally (typically 180°C, 15 h) (162), photochemically (UV or even visible illumination) (163), or thermally with the help of a Lewis-acid catalyst ($EtAlCl_2$, 100°C, 15 h) (159,164), the latter method severely limiting the choice of the terminal function. In order to avoid the formation of oxide, it is important to rigorously exclude moisture from the grafting solution. In practice, this is automatically warranted if a catalyst such as EtAlCl$_2$ is used. Also, when preparing mixed decyl/10-carboxydecyl monolayers from decene/undecylenic acid mixtures, the presence of a carboxylic acid has been found to be beneficial for preventing oxidation, which is attributed to water trapping in the form of inverse micelles in the decene solvent (165).

Here again, steric hindrance usually limits the substitution of surface hydrogen atoms to at most 50%. SiH groups are then left on the surface below the organic layer, and the concentration of alkyl chains in these layers is on the order of 3-4 10^{14} cm^{-2}, which makes SiH oxidation by water penetration through the layer very difficult, though not quite impossible, especially at high pH. Ageing of these surfaces in ambient air on a month time scale or in basic solution for a much shorter duration is indeed found to lead to some oxidation (166). It has been reported that starting from alkyne precursors instead of alkenes leads to a higher packing of the chains, hence an enhanced stabilization of the surface against oxidation is obtained (167).

Modifying the grafted layer: toward sensor applications

Once an organic layer has been grafted on a silicon surface, conventional chemical methods may be applied for modifying this layer. Starting from an alkyl layer is possible, though not easy given its poor reactivity. Strong oxidative treatments using an oxygen plasma (168,169) have been reported to lead to the creation of oxygenated functions (alcohols, carbonyl and carboxyl groups). Silicon oxidation may be prevented if the plasma is very mild (168). However, the oxidation reaction is hardly stereospecific (oxidation occurs at uncontrolled positions on the grafted chains), and avoiding simultaneous silicon oxidation is difficult. The anchoring of carboxylic acid functions by direct hydrosilylation of undecylenic acid (170) or of ethyl undecylenate, followed by hydrolysis of the ester (171), provides more convenient routes for the immobilization of carboxyl-terminated groups. These represent a better defined and more convenient starting point for chemical engineering of the surface using conventional organic synthesis methods.

The immobilization of chemical or biochemical species on a carboxyl-terminated surface is most often obtained by activating the carboxyl groups with a carbodiimide, enabling fast reaction of the activated terminations with a primary amine function. The method uses mild conditions (neutral pH, room temperature) and is not harmful for silicon. A better control of the activation is often obtained by adding N-hydroxysuccinimide to the activation solution (172). The activated species formed is then a succinimidyl ester, which can be handled safely before reacting with the amine. Alternately, alkene chains bearing a succinimidyl ester termination may be synthesized and grafted directly onto silicon by hydrosilylation (173). Any molecule bearing a primary amine linker may then be immobilized on the surface through a robust covalent amide bond (Fig. 5). It has been shown that these treatments, initially assessed on well-defined {111} silicon surfaces, can be transposed to the surface of porous silicon (174), and also of plasma-deposited amorphous hydrogenated silicon and silicon/carbon alloys (174,175), which present many advantages in terms of cost and flexibility. Especially, these materials can be deposited on any surface at any desired thickness, which has allowed for the design and optimization of sensors based on fluorescence, surface plasmon, and localized surface plasmon detection (175,176). Many different species, such as peptides, DNA oligomers, glycans, and various proteins, have been immobilized on such surfaces (175-178). The concentration of immobilized molecules can be controlled by starting from a layer where the carboxyl-terminated groups are diluted among methyl-terminated groups. Such surfaces may be obtained by using a grafting solution composed of carboxyl-terminated precursors diluted in decene (170). For sensor applications, non-specific adsorption of the target molecules on the hydrophobic surface is a problem, which may be avoided if the alkene chains are modified by attachment of hydrophilic ethyleneoxide oligomers at their end (177). Though the organic coating does not rigorously exclude silicon oxidation on the long term, the protection is sufficient to guarantee perfect stability, reproducibility and reusability of these sensors on a time scale of several months (175).

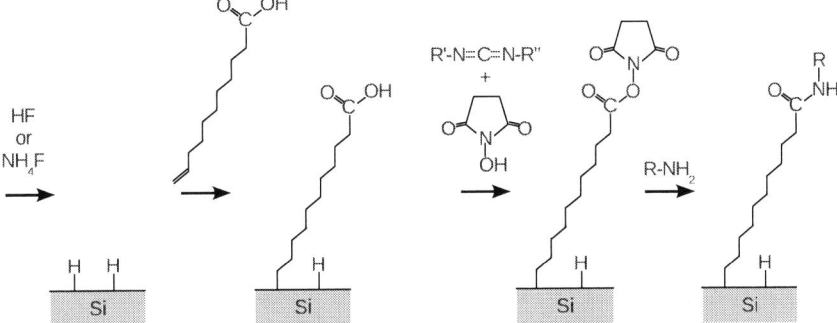

Figure 5. Typical scheme for the immobilization of (bio)molecules on the silicon surface.

Conclusion: Challenges for the Near Future

After more than 50 years investigation of the electrochemistry of silicon, we are still left with many fundamental unanswered questions. The electrochemistry of silicon in fluoride medium is especially intriguing. There is as yet no full agreement on the formation mechanism(s) of porous silicon. Addressing this issue would require also to take into account the many more recent observations on the formation of nanowire arrays (75,76), and the formation of other porous semiconductors (61). Another challenge is the understanding of the origin of the macromorphologies obtained for porous silica (131), which appear surprisingly different from those obtained for porous alumina and porous titania (129,130). Also, the nature of the mechanism at the origin of the electropolishing oscillations is still controversial. We do believe that progress on this point will require consideration of the voltammogram as a whole, and not only of that part where the oscillations are observed. We have made a step in that direction by developing a model for a "wet oxide", which correctly reproduces the current and the amount of oxide in region (B) (179). Extension to region (C) is presently under consideration.

The development of modified silicon surfaces for sensor applications is a field of wide applied interest. Long-term stabilization of the silicon surface against oxidation is a key challenge in that field. For some people, the issue appears hopeless, and diamond appears as a more promising material (180). We do think, and our recent results demonstrate, that silicon is still a good competitor in that field. Our results on fluorescence and plasmon sensors based on amorphous hydrogenated silicon/carbon alloys indicate a fair stability, sufficient for the development of such sensors. For field-effect sensors, a very hot subject in the context of sensors based on silicon nanowires (181), the issue is much more critical, because the oxidation, associated with the formation of silanol groups and interface states, may lead to an undesired sensitivity to the pH and to drifts in the sensor characteristics. An improved performance of the organic coatings against oxidation then appears as a major challenge for such applications. The grafting from alkynes instead of alkenes has been reported to represent a progress in that direction (167). A comparative assessment of the performances of the two approaches is still to be made. Other methods may be considered too: partial oxidation of the grafted chains has been reported to enhance their packing, plausibly offering a better protection against water penetration (168). Also, one may imagine the grafting of suitable precursors, that might be cross-linked after grafting.

Finally, silicon has been recently pointed out as a promising material for lithium battery anodes, as its theoretical capacity for lithium insertion exceeds that of graphite by one order of magnitude (182). The problem of its poor cyclability is becoming an issue of considerable interest. Using nanostructured forms of silicon may be a way to solve the problem (183). Our recent results indicate that it may also be circumvented by using suitable amorphous silicon-carbon alloys (184). There is little doubt that exciting developments in these directions will take place in the near future.

Acknowledgments

The author acknowledges the long-standing collaboration and friendship of F. Ozanam and P. Allongue. He is also indebted to all the members of the Electrochemistry and Thin Films group at Laboratoire de Physique de la Matière Condensée, and to the many collaborators, from inside or outside the laboratory, senior researchers, postdocs or PhD students, that he had the chance to work with over the years and who participated in the work on silicon reviewed here.

References

1. A. Uhlir Jr., *Bell Syst. Tech. J.*, **35**, 333 (1956).
2. D. R. Turner, *J. Electrochem. Soc.*, **105**, 402 (1958).
3. R. Memming and G. Schwandt, *Surf. Sci.*, **4**, 109 (1966).
4. R. Memming and G. Schwandt, *Surf. Sci.*, **5**, 97(1966).
5. A. Fujishima and K. Honda, *Nature* **238**, 37 (1972).
6. H. O. Finklea, Editor, *Semiconductor Electrodes*, Elsevier, Amsterdam, the Netherlands (1988).
7. R. L. Smith and S. D. Collins, *J. Appl. Phys.*, **71**, R1 (1992).
8. L. T. Canham, Editor, *Properties of Porous Silicon*, EMIS Datareviews No 18, The Institution of Electrical Engineers, London, United Kingdom (1997).
9. L. T. Canham, *Appl. Phys. Lett.*, **57**, 1046 (1990).
10. X. G. Zhang, *Electrochemistry of Silicon and its Oxide*, Kluwer/Plenum, New

York (2001).

11. V. Lehmann, *Electrochemistry of Silicon*, Wiley-VCH, Weinheim, Germany (2002).
12. A. Sassolas, B. D. Leca-Bouvier, and L. J. Blum, *Chem. Rev.*, **108**, 109 (2008).
13. R. Boukherroub, *Curr. Opin. Solid State Mater. Sci.*, **9**, 66 (2005).
14. S. Ciampi, J. B. Harper, and J. J. Gooding, *Chem. Soc. Rev.*, **39**, 2158 (2010).
15. D. Bélanger and J. Pinson, *Chem. Soc. Rev.* , **40**, 3995 (2011).
16. M. J. J. Theunissen, *J. Electrochem. Soc.*, **119**, 351 (1972).
17. P. F. Schmidt and W. Michel, *J. Electrochem. Soc.*, **104**, 230 (1957).
18. M. J. Madou, S. R. Morrison, and V. P. Bondarenko, *J. Electrochem. Soc.*, **135**, 229 (1988).
19. F. Gaspard, A. Halimaoui, and G. Sarrabayrouse, *Rev. Phys. Appl.*, **22**, 65 (1987).
20. J.-N. Chazalviel, *Surf. Sci.*, **88**, 204 (1979).
21. M. J. Madou, B. H. Loo, K. W. Frese, and S. R. Morrison, *Surf. Sci.*, **108**, 135 (1981).
22. J.-N. Chazalviel, *J. Electrochem. Soc.*, **127**, 1822 (1980).
23. A. Q. Contractor, M. Szklarczyk, and J. O'M. Bockris, *J. Electroanal. Chem.*, **157**, 175 (1983).
24. C. U. Maier, M. Specht, and G. Bilger, *Intl. J. Hydrogen Energy*, **21**, 859 (1996).
25. H. S. Abdel-Samad, M. A. Amin, J.-N. Chazalviel, F. Ozanam, and P. Allongue, *Electrochim. Acta*, **49**, 4577 (2004).
26. K. C. Mandal, F. Ozanam, and J.-N. Chazalviel, *Appl. Phys. Lett.*, **57**, 2788 (1990).
27. P. de Mierry, A. Etcheberry, R. Rizk, P. Etchegoin, and M. Aucouturier, *J. Electrochem. Soc.*, **141**, 1539 (1994).
28. A. Belaïdi, J.-N. Chazalviel, F. Ozanam, O. Gorochov, A. Chari, B. Fotouhi, and M. Etman, *J. Electroanal. Chem.*, **444**, 55 (1998).
29. P. de Mierry, D. Ballutaud, M. Aucouturier, and A. Etcheberry, *J. Electrochem. Soc.*, **137**, 2966 (1990).
30. R. A. Wind, H. Jones, M. J. Little, and M. A. Hines, *J. Phys. Chem. B*, **106**, 1557 (2002).
31. R. L. Smith, B. Kloeck, N. de Rooij, and S. D. Collins, *J. Electroanal. Chem.*, **238**, 103 (1987).
32. E. D. Palik, V. M. Bermudez, and O. J. Glembocki, *J. Electrochem. Soc.*, **132**, 871 (1985).
33. P. Allongue, V. Costa-Kieling, and H. Gerischer, *J. Electrochem. Soc.*, **140**, 1009 (1993).
34. P. Allongue, V. Costa-Kieling, and H. Gerischer, *J. Electrochem. Soc.*, **140**, 1018 (1993).
35. H. G. G. Philipsen and J. J. Kelly, *J. Phys. Chem. B*, **109**, 17245 (2005).
36. N. J. Harrick and K. H. Beckmann, in *Characterization of Solid Surfaces*, P. F. Kane and G. R. Larrabee, Editors, pp. 215-245, Plenum, New York (1974).
37. H. Ubara, T. Imura, and A. Hiraki, *Solid State Commun.*, **50**, 673 (1984).
38. A. Tardella and J.-N. Chazalviel, *Appl. Phys. Lett.*, **47**, 334 (1985).
39. E. Yablonovitch, D. L. Allara, C. C. Chang, T. Gmitter, and T. B. Bright, *Phys. Rev. Lett.*, **57**, 249 (1986).
40. V. A. Burrows, Y. J. Chabal, G. S. Higashi, K. Raghavachari, and S. B. Christman, *Appl. Phys. Lett.*, **53**, 998 (1988).
41. G. W. Trucks, K. Raghavachari, G. S. Higashi, and Y. J. Chabal, *Phys. Rev. Lett.*,

65, 504 (1990).

42. G. S. Higashi, Y. J. Chabal, G. W. Trucks, and K. Raghavachari, *Appl. Phys. Lett.*, **56**, 656 (1990).

43. P. Allongue, V. Kieling, and H. Gerischer, *Electrochim. Acta*, **40**, 1353 (1995).

44. E. Landemark, C. J. Karlsson, and R. I. G. Uhrberg, *Phys. Rev. B*, **44**, 1950 (1991).

45. H. Yasufuku, K. Meguro, S.-I. Akatsuka, H. Setoyama, S. Kera, Y. Azuma, K. K. Okudaira, S. Hasegawa, Y. Harada, and N. Ueno, *Jpn. J. Appl. Phys.*, **39**, 1706 (2000).

46. M. L. Munford, R. Cortès, and P. Allongue, *Sens. Mater.*, **13**, 259 (2001).

47. H. Gerischer and M. Lübke, *Ber. Bunsenges. Phys. Chem.*, **91**, 394 (1987).

48. A. E. Gershinskii and L. V. Mironova, *Sov. Electrochem.*, **25**, 1224 (1989).

49. M. J. Eddowes, *J. Electroanal. Chem.*, **280**, 297 (1990).

50. J.-N. Chazalviel, M. Etman, and F. Ozanam, *J. Electroanal. Chem.*, **297**, 533 (1991).

51. J.-N. Chazalviel, in *Porous Silicon Science and Technology*, J.-C. Vial and J. Derrien, Editors, pp. 17-32, Springer, Berlin (1995).

52. F. Gaspard, A. Bsiesy, M. Ligeon, F. Muller, and R. Hérino, *J. Electrochem. Soc.*, **136**, 3043 (1989).

53. P. C. Searson and X.G. Zhang, *J. Electrochem. Soc.*, **137**, 2539 (1990).

54. F. Ozanam, J.-N. Chazalviel, A. Radi, and M. Etman, *J. Electrochem. Soc.*, **139**, 2491 (1992).

55. T. Dittrich, V. Y. Timoshenko, and J. Rappich, *Appl. Phys. Lett.*, **72**, 1635 (1998).

56. M. Matsumura and S. R. Morrison, *J. Electroanal. Chem.*, **147**, 157 (1983).

57. H. Gerischer and M. Lübke, *J. Electrochem. Soc.*, **135**, 2782 (1988).

58. A. E. Gershinskii, L. V. Mironova, and E. I. Cherepov, *Phys. Status Solidi (a)*, **38**, 369 (1976).

59. F. Bensliman, N. Mizuta, and M. Matsumura, *J. Electroanal. Chem.*, **568**, 353 (2004).

60. A. Venkateswara Rao, F. Ozanam, and J.-N. Chazalviel, *J. Electrochem. Soc.*, **138**, 153 (1991).

61. see, e.g., the proceedings of the conferences series *Porous Semiconductors Science and Technology*: Valencia (1998), Madrid (2000), Tenerife (2002), Valencia (2004), Barcelona (2006), Mallorca (2008), Valencia (2010), Malaga (2012).

62. R. B. Wehrspohn, F. Ozanam, and J.-N. Chazalviel, *J. Electrochem. Soc.*, **146**, 3309 (1999).

63. C. Lévy-Clément, A. Lagoubi, D. Ballutaud, F. Ozanam, J.-N. Chazalviel, and M. Neumann-Spallart, *Appl. Surf. Sci.*, **65-66**, 408 (1993).

64. V. Lehmann and H. Föll, *J. Electrochem. Soc.* **137**, 653 (1990).

65. V. Lehmann and S. Rönnebeck, *J. Electrochem. Soc.*, **146**, 2968 (1999).

66. J.-N. Chazalviel and F. Ozanam, in Ref. (8), chap. 1.8.

67. E. M. Media, J.-N. Chazalviel, F. Ozanam, and R. Outemzabet, *Phys. Status Solidi (c)*, **8**, 1727 (2011).

68. H. Okayama, K. Fukami, R. Plugaru, T. Sakka, and Y. H. Ogata, *J. Electrochem. Soc.*, **157**, D54 (2010) and references therein.

69. N. Tit, Z. H. Yamani, G. Pizzi, and M. Virgilio, *Phys. Status Solidi (c)*, **9**, 1458 (2012) and references therein.

70. R. B. Wehrspohn, J.-N. Chazalviel, F. Ozanam, and I. Solomon, *Eur. Phys. J. B*,

8, 179 (1999).

71. see, e.g., B. Gelloz, T. Shibata, and N. Koshida, *Appl. Phys. Lett.*, **89**, 191103 (2006).
72. K. W. Kolasinski, *Curr. Opin. Solid State Mater. Sci.*, **9**, 73 (2005).
73. I. Solomon, K. Rerbal, J.-N. Chazalviel, F. Ozanam, and R. Cortès, *J. Appl. Phys.*, **103**, 083108 (2008).
74. K. Tsujino and M. Matsumura, *Adv. Mater.*, **17**, 1045 (2005).
75. K. Q. Peng, J. J. Hu, Y. J. Yan, Y. Wu, H. Fang, Y. Xu, S. T. Lee, and J. Zhu, *Adv. Funct. Mater.*, **16**, 387 (2006).
76. R. Douani, G. Piret, T. Hadjersi, J.-N. Chazalviel, and I. Solomon, *Thin Solid Films*, **519**, 5383 (2011).
77. X. G. Zhang, *J. Electrochem. Soc.*, **138**, 3750 (1991).
78. M. H. Al Rifai, M. Christophersen, S. Ottow, J. Carstensen, and H. Föll, *J. Electrochem. Soc.*, **147**, 627 (2000).
79. V. Lehmann, R. Stengl, and A. Luigart, *Mat. Sci. Engin. B*, **69-70**, 11 (2000).
80. M. I. J. Beale, J. D. Benjamin, M. J. Uren, N. G. Chew, and A. G. Cullis, *J. Cryst. Growth*, **73**, 622 (1985).
81. A. Valance, *Phys. Rev. B*, **55**, 9706 (1997).
82. V. Lehmann and U. Gösele, *Appl. Phys. Lett.*, **58**, 856 (1991).
83. J. Carstensen, M. Christophersen, and H. Föll, *Mat. Sci. Engin. B*, **69-70**, 23 (2000).
84. J. W. Corbett, D. I. Shereshevskii, and I. V. Verner, *Phys. Status Solidi (a)*, **147**, 81 (1995).
85. P. Allongue, C. Henry de Villeneuve, L. Pinsard, and M. C. Bernard, *Appl. Phys. Lett.*, **67**, 941 (1995).
86. E. S. Kooj and D. Vanmaekelbergh, *J. Electrochem. Soc.*, **144**, 1296 (1997).
87. D. M. Soares, M. C. dos Santos, and O. Teschke, *Chem. Phys. Lett.*, **242**, 202 (1995).
88. J.-N. Chazalviel, F. Ozanam, N. Gabouze, S. Fellah, and R. B. Wehrspohn, *J. Electrochem. Soc.*, **149**, C511 (2002).
89. J.-N. Chazalviel and F. Ozanam, in *Ordered Porous Nanostructures and Applications*, R. B. Wehrspohn, Editor, Plenum, New York (2005).
90. X. G. Zhang, *J. Electrochem. Soc.*, **151**, C69 (2004).
91. H. H. Hassan, J. L. Sculfort, M. Etman, F. Ozanam, and J.-N. Chazalviel, *J. Electroanal. Chem.*, **380**, 55 (1995).
92. S. Cattarin, I. Frateur, M. Musiani, and B. Tribollet, *J. Electrochem. Soc.*, **147**, 3277 (2000).
93. J. E. A. M. van den Meerakker and M. R. L. Mellier, *J. Electrochem. Soc.*, **148**, G166 (2001).
94. A. Belaïdi, M. Safi, F. Ozanam, J.-N. Chazalviel, and O. Gorochov, *J. Electrochem. Soc.*, **146**, 2659 (1999).
95. C. da Fonseca, F. Ozanam, and J.-N. Chazalviel, *Surf. Sci.*, **365**, 1 (1996).
96. V. Lehmann, *J. Electrochem. Soc.*, **140**, 2836 (1993).
97. R. Outemzabet, M. Cherkaoui, N. Gabouze, F. Ozanam, N. Kesri, and J.-N. Chazalviel, *J. Electrochem. Soc.*, **153**, C108 (2006).
98. J. Rappich, V. Y. Timoshenko, and T. Dittrich, *Mat. Res. Soc. Symp. Proc.*, **448**, 51 (1997).
99. J. Rappich H. Jungblut, M. Aggour, and H. J. Lewerenz, *J. Electrochem. Soc.*,

141, L99 (1994).

100. F. Ozanam, C. da Fonseca, A. Venkateswara Rao, and J.-N. Chazalviel, *Appl. Spec.*, **51**, 519 (1997).
101. E. Peiner and A. Schlachetzki, *J. Electrochem. Soc.*, **139**, 552 (1992).
102. H. Gerischer and M. Lübke, *Ber. Bunsenges. Phys. Chem.*, **92**, 573 (1988).
103. F. Ozanam, J.-N. Chazalviel, A. Radi, and M. Etman, *Ber Bunsenges Phys. Chem.*, **95**, 98 (1991).
104. J.-N. Chazalviel and F. Ozanam, *J. Electrochem. Soc.*, **139**, 2501 (1992).
105. J.-N. Chazalviel, F. Ozanam, M. Etman, F. Paolucci, L. M. Peter, and J. Stumper, *J. Electroanal. Chem.*, **327**, 343 (1992).
106. F. Ozanam and J.-N. Chazalviel, *J. Electron Spec. Relat. Phenom.*, **64/65**, 395 (1993).
107. H. J. Lewerenz and M. Aggour, *J. Electroanal. Chem.*, **351**, 159 (1993).
108. F. Ozanam, N. Blanchard, and J.-N. Chazalviel, *Electrochim. Acta*, **38**, 1627 (1993).
109. V. Lehmann, *J. Electrochem. Soc.*, **143**, 1313 (1996).
110. J. Rappich, V. Y. Timoshenko, and T. Dittrich, *Ber. Bunsenges. Phys. Chem.*, **101**, 139 (1997).
111. S. Cattarin, J.-N. Chazalviel, C. da Fonseca, F. Ozanam, L. M. Peter, O. Schlichthorl, and J. Stumper, *J. Electrochem. Soc.*, **145**, 498 (1998).
112. J.-N. Chazalviel, C. da Fonseca, and F. Ozanam, *J. Electrochem. Soc.*, **145**, 964 (1998).
113. M. Bailes, S. Böhm, L. M. Peter, D. J. Riley, and R. Greef, *Electrochim. Acta*, **43**, 1757 (1998).
114. J. Carstensen, R. Prange, G. S. Popkirov, and H. Föll, *Appl. Phys. A*, **67**, 459 (1998).
115. O. Nast, S. Rauscher, H. Jungblut, and H. J. Lewerenz,, *J. Electroanal. Chem.*, **442**, 169 (1998).
116. S. Cattarin, F. Decker, D. Dini, and B. Margesin, *J. Electroanal. Chem.*, **474**, 182 (1999).
117. J. Carstensen, R. Prange, and H. Föll, *J. Electrochem. Soc.*, **146**, 1134 (1999).
118. J. Grzanna, H. Jungblut, and H. J. Lewerenz, *J. Electroanal. Chem.*, **486**, 181 (2000).
119. J. Grzanna, H. Jungblut, and H. J. Lewerenz, *J. Electroanal. Chem.*, **486**, 190 (2000).
120. V. Parkhutik, E. Matveeva, R. Perez, J. Alamo, and D. Beltrán, *Mat. Sci. Engin. B*, **69-70**, 553 (2000).
121. E. Foca, J. Carstensen, and H. Föll, *J. Electroanal. Chem.*, **603**, 175 (2007).
122. I. Miethe, V. Garcia-Morales, and K. Krischer, *Phys. Rev. Lett.*, **102**, 194101 (2009).
123. J.-N. Chazalviel and F. Ozanam, *Electrochim. Acta*, **55**, 656 (2010).
124. I. Miethe and K. Krischer, *J. Electroanal. Chem.*, **666**, 1 (2012).
125. M. Lharch, J.-N. Chazalviel, F. Ozanam, M. Aggour, and R. B. Wehrspohn, *Phys. Stat. Sol. (a)*, **197**, 39 (2003).
126. S. Frey, B. Grésillon, F. Ozanam, J.-N. Chazalviel, J. Carstensen, H. Föll, and R. B. Wehrspohn, *Electrochem. Solid-State Lett.*, **8**, B25 (2005).
127. S. Frey, S. Keipert, J.-N. Chazalviel, F. Ozanam, J. Carstensen, and H. Föll, *Phys. Status Solidi (a)*, **204**, 1250 (2007).
128. J.-N. Chazalviel and F. Ozanam, *Materials*, **4**, 825 (2011).

129. J. M. Macak, K. Sirotna, and P. Schmuki, *Electrochim. Acta*, **50**, 3679 (2005).
130. O. Jessensky, F. Müller, and U. Gösele, *Appl. Phys. Lett.*, **72**, 1173 (1998).
131. M. A. Amin, S. Frey, F. Ozanam, and J.-N. Chazalviel, *Electrochim. Acta*, **53**, 4485 (2008).
132. J.-N. Chazalviel, R. Cortès, F. Maroun, and F. Ozanam, *Phys. Status Solidi (a)*, **206**, 1229 (2009).
133. J.-N. Chazalviel and F. Ozanam, *ECS Trans.*, **25(27)**, 131 (2010).
134. E. A. Ponomarev and C. Lévy-Clément, *J. Porous Mater.*, **7**, 51 (2000).
135. M. Christophersen, J. Carstensen, S. Rönnebeck, C. Jäger, and H. Föll, *J. Electrochem. Soc.*, **148**, E267 (2001).
136. M. Yang, N. K. Shrestha, and P. Schmuki, *Electrochem. Solid-State Lett.*, **13**, C25 (2010).
137. E. K. Propst and P. A. Kohl, *J. Electrochem. Soc.*, **141**, 1006 (1994).
138. K. D. Legg, A. B. Ellis, J. M. Bolts, and M. S. Wrighton, *Proc. Natl. Acad. Sci.*, **74**, 4116 (1977).
139. R. E. Malpas, K. Itaya, and A. J. Bard, *J. Am. Chem. Soc.*, **103**, 1622 (1981).
140. J.-N. Chazalviel and T.B. Truong, *J. Am. Chem. Soc.*, **103**, 7447 (1981).
141. J.-N. Chazalviel, *J. Electrochem. Soc.*, **129**, 963 (1982).
142. M. L. Rosenbluth and N. S. Lewis, *J. Am. Chem. Soc.*, **108**, 4689 (1986).
143. J. F. Gibbons, G. W. Cogan, C. M. Gronet, and N. S. Lewis, *Appl. Phys. Lett.*, **45**, 1095 (1984).
144. J.-N. Chazalviel, *J. Electroanal. Chem.*, **233**, 37 (1987).
145. M. Warntjes, C. Vieillard, F. Ozanam, and J.-N. Chazalviel, *J. Electrochem. Soc.*, **142**, 4138 (1995).
146. R. Boukherroub, S. Morin, P. Sharpe, D. D. M. Wayner, and P. Allongue, *Langmuir*, **16**, 7429 (2000).
147. J. Pinson and F. Podvorica, *Chem. Soc. Rev.*, **34**, 429 (2005).
148. C. Henry de Villeneuve, J. Pinson, M.-C. Bernard, and P. Allongue, *J. Phys. Chem. B*, **101**, 2415 (1997).
149. P. Allongue, C. Henry de Villeneuve, G. Cherouvrier, R. Cortès, and M.-C. Bernard, *J. Electroanal. Chem.*, **550-551**, 161 (2003).
150. F. Aït El Hadj, A. Amiar, M. Cherkaoui, J.-N. Chazalviel, and F. Ozanam, *Electrochim. Acta*, **70**, 318 (2012).
151. S. Fellah, A. Teyssot, F. Ozanam, J.-N. Chazalviel, J. Vigneron, and A. Etcheberry, *Langmuir*, **18**, 5851 (2002).
152. A. Fidélis, F. Ozanam, and J.-N. Chazalviel, *Surf. Sci.*, **444**, L7 (2000).
153. D. Dusciac, V. Brizé, J.-N. Chazalviel, Y.-F. Lai, H. Roussel, S. Blonkowski, R. Schafranek, A. Klein, C. Henry de Villeneuve, P. Allongue, F. Ozanam, and C. Dubourdieu, *Chem. Mater.*, **24**, 3135 (2012).
154. S. Fellah, F. Ozanam, J.-N. Chazalviel, J. Vigneron, A. Etcheberry, and M. Stchakovsky, *J. Phys. Chem. B*, **110**, 1665 (2006).
155. S. Fellah, A. Amiar, F. Ozanam, J.-N. Chazalviel, J. Vigneron, A. Etcheberry, and M. Stchakovsky, *J. Phys. Chem. B*, **111**, 1310 (2007).
156. C. Gurtner, A. W. Wun, and M. J. Sailor, *Angew. Chem. Intl. Ed.*, **38**, 1966 (1999).
157. E. G. Robins, M. P. Stewart, and J. M. Buriak, *J. Chem. Soc.-Chem. Commun.*, 2479 (1999).
158. N. Y. Kim and P. E. Laibinis, *J. Am. Chem. Soc.*, **120**, 4516 (1998).
159. R. Boukherroub, S. Morin, F. Bensebaa, and D. D. M. Wayner, *Langmuir*, **15**,

3831 (1999).

160. S. Fellah, R. Boukherroub, F. Ozanam, and J.-N. Chazalviel, *Langmuir*, **20**, 6359 (2004).

161. A. Bansal, X. Li, I. Lauermann, N. S. Lewis, S. I. Yi, and W. H. Weinberg, *J. Am. Chem. Soc.*, **118**, 7225 (1996).

162. A. B. Sieval, V. Vleeming, H. Zuilhof, and E. J. R. Sudhölter, *Langmuir*, **15**, 8288 (1999).

163. R. L. Cicero, M. R. Linford, and C. E. D. Chidsey, *Langmuir*, **16**, 5688 (2000).

164. J. M. Buriak and M. J. Allen, *J. Am. Chem. Soc.*, **120**, 1339 (1998).

165. D. Aureau, thèse de l'Ecole Polytechnique (2008). http://hal.archives-ouvertes.fr/docs/00/50/10/40/PDF/version_finale_AUREAU.pdf.

166. D. Aureau, J. Rappich, A. Moraillon, P. Allongue, F. Ozanam, and J.-N. Chazalviel, *J. Electroanal. Chem.*, **646**, 33 (2010).

167. G. F. Cerofolini, C. Galati, S. Reina, and L. Renna, *Surf. Interf. Anal.*, **38**, 126 (2006).

168. D. Aureau, W. Morscheidt, A. Etcheberry, J. Vigneron, F. Ozanam, P. Allongue, and J.-N. Chazalviel, *J. Phys. Chem. C*, **113**, 14418 (2009).

169. M. Rosso, M. Giesbers, K. Schroën, and H. Zuilhof, *Langmuir*, **26**, 866 (2010).

170. A. Faucheux, A. C. Gouget-Laemmel, C. Henry de Villeneuve, R. Boukherroub, F. Ozanam, P. Allongue, and J.-N. Chazalviel, *Langmuir* ,**22**, 153 (2006).

171. A. B. Sieval, A. L. Demirel, J. W. M. Nissink, M. R. Linford, J. H. van der Maas, W. H. de Jeu, H. Zuilhof, and E. J. R. Sudhölter, *Langmuir*, **14**, 1759 (1998).

172. J. V. Staros, R. W. Wright, and D. M. Swingle, *Anal. Biochem.*, **156**, 220 (1986).

173. J. T. C. Wojtyk, K. A. Morin, R. Boukherroub, and D. D. M. Wayner, *Langmuir*, **18**, 6081 (2002).

174. R. Boukherroub, J. T. C. Wojtyk, D. D. M. Wayner, and D. J. Lockwood, *J. Electrochem. Soc.*, **149**, H59 (2002).

175. L. Touahir, P. Allongue, D. Aureau, R. Boukherroub, J.-N. Chazalviel, E. Galopin, A. C. Gouget-Laemmel, C. Henry de Villeneuve, A. Moraillon, J. Niedziółka-Jönsson, F. Ozanam, J. Salvador Andresa, S. Sam, I. Solomon, and S. Szunerits, *Bioelectrochem.*, **80**, 17 (2010).

176. J.-N. Chazalviel, P. Allongue, A. C. Gouget-Laemmel, C. Henry de Villeneuve, A. Moraillon, and F. Ozanam, *Sci. Adv. Mater.*, **3**, 332 (2011).

177. E. Perez, K. Lhalil, C. Rougeau, A. Moraillon, J.-N. Chazalviel, F. Ozanam, and A. C. Gouget-Laemmel, *Langmuir*, **28**, 14654 (2012).

178. A. C. Gouget-Laemmel, J. Yang, M. A. Lodhi, A. Siriwardena, D. Aureau, R. Boukherroub, J.-N. Chazalviel, F. Ozanam, and S. Szunerits, *J. Phys. Chem. C*, **117**, 368 (2013).

179. R. Cheggou, A. Kadoun, N. Gabouze, F. Ozanam, and J.-N. Chazalviel, *Electrochim. Acta*, **54**, 3053 (2009).

180. R. J. Hamers, J. E. Butler, T. Lasseter, B. M. Nichols, J. N. Russell Jr., K.-Y. Tse, and W. Yang, *Diamond Relat. Mater.*, **14**, 661 (2005).

181. R. M. Penner, *Ann. Rev. Anal. Chem.*, **5**, 461 (2012).

182. M. N. Obrovac, L. Christensen, D. B. Le, and J. R. Dahn, *J. Electrochem. Soc.*, **154**, A849 (2007).

183. W.-J. Zhang, *J. Power Sources*, **196**, 13 (2011).

184. M. Rosso, L. Touahir, A. Cheriet, I. Solomon, J.-N. Chazalviel, F. Ozanam, and N. Gabouze, International Patent Application PCT/FR2012/051161, published as WO/2012/160315 (29/11/2012).

CHAPTER 2

ELECTROCHEMICAL AND SEMICONDUCTOR DEVICES

24

Semiconductor Nanostructures for Antireflection Coatings, Transparent Contacts, Junctionless Thermoelectrics and Li-ion Batteries

C. Glynn[1,2], M. Osiak[1], W. McSweeney[1,3], O. Lotty[5], K. Jones[3], H. Geaney[1,2], E. Quiroga-González[6], J. D. Holmes[2,5,7], and C. O'Dwyer[1,2,4]

[1] Department of Chemistry, University College Cork, Cork, Ireland
[2] Micro- and Nanoelectronics Centre, Tyndall National Institute, Lee Maltings, Cork, Ireland
[3] Department of Physics & Energy, University of Limerick, Limerick, Ireland
[4] Materials and Surface Science Institute, University of Limerick, Limerick, Ireland
[5] Materials Chemistry and Analysis Group, Department of Chemistry University College Cork, Cork, Ireland
[6] Institute for Materials Science, Christian-Albrechts Universität, Kiel, Germany
[7] Centre for Research on Adaptive Nanostructures and Nanodevices (CRANN), Trinity College Dublin, Dublin 2, Ireland

Porous semiconductors structured top-down by electrochemical means, and from bottom-up growth of arrays and arrangements of nanoscale structures, are shown to be amenable to a range of useful thermal, optical, electrical and electrochemical properties. This paper summarises recent investigations of the electrochemical, electrical, optical, thermal and structural properties of porous semiconductors such as Si, In_2O_3, SnO_2 and ITO, and dispersions, arrays and arrangements of nanoscale structures of each of these materials. We summarize the property-inspired application of such structurally engineered arrangements and morphologies of these materials for antireflection coatings, broadband absorbers, transparent contacts to LEDs that improve transmission, electrical contact and external quantum efficiency. Additionally the possibility of thermoelectric performance through structure-mediated variation in thermal resistance and phonon scattering without a p-n junction is shown through phonon engineering in roughened nanowires. Lastly, we show that bulk crystals and nanowires of p- and n-type doped Si are promising for use as anodes in Li-ion batteries.

Introduction

Porous semiconductors continue to receive considerable attention because of functionality afforded by random or ordered structuring on the nano or mesoscale. For instance, a wide range of mono- and compound semiconductors can be rendered porous through electroless or electrochemical etching (1-26), and such top down approaches allow a high degree of control over porosity formation. Additionally, the resulting skeleton formed through these means allows the possibility for arrays or arrangements of nanostructured materials such as nanowires (NWs), mesoporous materials, nanorods, photonic crystals and many other structural forms. When the sizes of the remaining semiconductors is reduced, photonic and phononic confinements effects can allow for rich optical and thermal effects not possible in the bulk materials. Additionally, porous materials can often alleviate structural changes when applied as electrodes in Li-ion batteries and other phase change or structure converting processes. Silicon has maintained a strong fundamental and applied research value, and recently has become one of the most significant materials, when structured on the nanoscale, for Li-ion battery anodes (27-29) and in thermoelectric materials evaluation (30-32).

Transparent conducting oxides are also of prime importance for electronic and photonic devices. Their major limitation is the trade-off between their conductivity and transparency (33), which can be overcome to some extent through controlled top-down or bottom up porosity formation, or effective porosity formation through nanostructured growth methods. In developing techniques to fabricate homogeneous one-dimensional (1D) nanostructures, researchers have sought to control shape, aspect ratio and crystalline arrangement (34-35), and recent improvements in synthetic methods (36-38) have led to the direct integration of functional nanostructures into nanoscale devices. Indium tin oxide (ITO) (39,40) is the most important transparent conducting oxide and is therefore used in a wide range of applications. However, to date it has enjoyed only limited success as an ohmic contact for light-emitting devices (LEDs) due to high resistivities and an unacceptable trade-off between electrical and optical characteristics. Moreover, the limited availability of materials with suitable refractive indices has prevented the implementation of optical components with high performance. Hence, the possibility of forming complex, multilevel branched structures (41) with optimized optical properties in a single-step, bottom-up growth regime would be a significant advance given their compatibility with optoelectronic device architectures. The development of high mobility oxides, or dispersion of metallic NWs that offer percolating conductivity and porosity to mimic a transparent conductive thin film, are important for charge storage devices such as batteries and pseudocapacitors, planar and flexible electronic and photonic devices.

Rechargeable batteries have been critical for enabling portable consumer electronics and are beginning to be used in electric vehicles. They are also becoming an attractive option for large-scale stationary energy storage (42-47). For mobile applications, high energy (per weight and volume) is the most important parameter since it determines the usage time per charge. For stationary applications, cost is the most important design parameter, and high energy batteries could help reduce the cost per unit of stored energy (48,49).

The continued development of porous structures for these reasons also has translational application to thermoelectrics and paradigms for energy harvesting from waste heat on the nanoscale in electronic devices, particularly LEDs and other high power devices, and the structuring of the porous semiconductors can tunable alter the thermal resistance of materials, allowing thermoelectric voltage generation without requiring a p-n junction. These types of materials and strategies to induced controlled porosity for a variety of functions are summarized for a series of common electronic and transparent oxide-type semiconductors.

Experimental

Molecular beam epitaxial growth of In_2O_3, SnO_2 and ITO nanodots and NWs

All surfaces were cleaned using a standard RCA process where the sample was immersed in a H_2O_2:NH_4OH:H_2O (1:1:1) solution at 80°C for 30 min. After rinsing in deionised water, a second treatment was performed in a H_2O_2:HCl:H_2O (1:1:5) solution with subsequent rinsing in deionized water. Through this process metallic and organic contamination is removed. In the latter step, the surface is oxidized so that a thin (~7 nm) and clean SiO_2 layer forms at the surface. Dipping the sample in HF shortly before introducing it into the evaporator removes the oxide layer and passivates the surface with hydrogen. For evaporation of the In and Sn sources, a home-built MBE high vacuum chamber with two distinct effusion cells for In and Sn together with an electron-beam evaporator, was designed in cooperation with MBE-Komponenten GmbH, Germany, with calibrated growth rates. During NW growth the sample surface was annealed at temperatures in the range 300°-650 °C. The In:Sn (90:10) was evaporated at growth rates in the range 0.02-0.2 nm s^{-1} up to maximum temperature for In and Sn of T_{In} = 835 °C, T_{Sn} = 1000 °C, respectively.

For In_2O_3 and SnO_2 nanodot dispersions, uniform layers of In and Sn metal were grown on respective substrates by depositing at a rate of 0.1 Å s^{-1} at a substrate temperature of 400 °C, with precise control over the nominal thickness, which was separately calibrated using a quartz crystal monitor. The nanodots then results from dewetting and oxidative crystallization.

Growth of Si/SiGe multi-quantum well devices

The n-i-p$^+$ Si/SiGe multi quantum well (MQW) structures were grown by molecular beam epitaxy (MBE). On an 80 nm undoped Si buffer grown at 700 °C on Si(001), a 15-period Si/Si$_{0.77}$Ge$_{0.23}$ (4.1 nm/3.9 nm) quantum well structure was deposited at 625 °C. The entire structure was capped with 42 nm B-doped (5 × 10^{18} cm^{-3}) Si to saturate SiGe dangling bonds. After an additional 10 nm Si top layer grown at 625 °C, a capping layer of 30 nm p$^+$-Si (p-doped) was grown at 700 °C. The back contact was a Ti (50 nm diffusion barrier)/Pt (50 nm)/Au (100 nm) multilayer ohmic contact. Several structures were top-contacted with a Ni/Cr bilayer deposited by sputtering.

Metal-assisted electroless etching of Si NW layers

Layers of p-type and n-type Si NWs, were fabricated by metal-assisted chemical (MAC) etching of 200 mm diameter lightly p-doped silicon (100) wafers (680 μm thickness) and highly doped n-type wafers, both with a native oxide layer (~2-5 nm thick). Substrates were immersed for two hours in a heated solution of 10% HF containing 0.04 M AgNO$_3$ and maintained at 50 °C using a thermostated water bath. Upon removal from the etching bath, samples were washed copiously with deionised water and then treated with concentrated nitric acid to remove unwanted silver deposition. The length of the NWs was controlled by the etching time and etchant concentration leaving a uniform, effective porous silicon layer. The remaining skeleton formed a vertical array of NWs ranging from 80 – 200 nm in diameter.

Optical, PL/EL and Raman spectroscopy measurements

Optical excitation came from a 6 W Argon ion (Ar$^+$) laser (Spectra Physics *Stabilite* 2017) operating at 514.5 nm. Luminescence was collected through a SPEX 1680 double monochromator into a cooled North Coast EO-817L Ge detector. For photoluminescence (PL) and electroluminescence (EL) measurements, a pulsed excitation optical signal with a fixed frequency of 590 Hz and an electrical signal pulsed at 570 Hz, respectively, were used and modulated by a mechanical chopper. The output from the chopper was fed through a lock-in amplifier measuring the output signal of the Ge detector sensitive in a spectral range 0.8-1.7 µm. Reflectance measurements were carried out in a Bruker FT-IR spectrometer IFS66/V. Different configurations of beam splitters, detectors and sources were used to cover the spectral range from a far infrared (10 cm^{-1}) to near infrared and visible ranges. For angular resolved measurements, a NIR512 Ocean Optics spectrometer was used as a detector in a home-built reflectance/transmittance setup.

Comparative photo- and electroluminescence (PL/EL) measurements were performed on Si/SiGe MQW LEDs grown by MBE. To compare the optical transmission, an ITO NW layer and a 30 nm NiCr metal contact were deposited on two separate samples on the same substrate. As a reference, the photoluminescence spectrum of an uncoated MQW structure was recorded under identical conditions.

Raman spectroscopy was conducted using a Dilor XY Labram spectrometer equipped with an Olympus BX40 confocal microscope and Renishaw InVia Raman spectrometer using a RenCam CCD camera. Excitation was provided by 514 nm ArHe 10 mW green laser with a maximum of 0.512 mW incident power. The spectra were collected with a Peltier cooled CCD detector. The incident power of the laser was adjusted using calibrated filters.

Structural, compositional and morphological characterization

All materials and samples were analysed using scanning and transmission electron microscopy (SEM/TEM). SEM was performed using both Hitachi S4800 and SU70 instruments equipped with an Oxford-50 mm^2 X-Max detector for energy dispersive X-ray analysis. TEM analysis was conducted with a JEOL JEM-2100F field emission microscope operating at 200 kV, equipped with a Gatan Ultrascan CCD camera and EDAX Genesis EDS detector for atomic resolution crystal structure and composition examination. For ultra-high resolution morphology, STEM and high angle annular dark field (HAADF) couple with electron diffraction was performed. X-ray diffraction (XRD) spectra were made using Cu Kα radiation with a Philips PanAlytical diffractometer in Bragg-Bretano geometry.

X-ray photoelectron spectroscopy was acquired using a Kratos Axis 165 monochromatized X-ray photoelectron spectrometer equipped with a dual anode (Mg/Al) source. Survey spectra were captured at as pass energy of 100 eV, step size of 1 eV, and dwell time of 50 ms. The core level spectra were an average of 10 scans captured at a PE of 25 eV, step size of 0.05 eV, and dwell time of 100 ms. The spectra were corrected for charge shift to the C 1s line at a binding energy of 284.9 eV. A Shirley background correction was employed, and the peaks were fitted to Voigt profiles.

Electrochemical analysis in Li-ion battery cells

Cyclic voltammetry and galvanostatic measurements were carried out using either 3-electrode cells or 2-electrode coin cells using a Multi Autolab 101 potentiostat and a BioLogic VSP potentio/galvanostat. All potentials, unless otherwise stated, are relative to Li$^+$/Li. Custom built Swagelok-type cells, split coin cells were used with counter and active material electrodes separated by a polypropylene or glass wool separators soaked in 1 mol dm^{-3} solution of LiPF$_6$ in EC:DMC at a 50:50 v/v ratio.

The electrodes were cycled at a scan rate of 0.5 mV/s in cyclic voltammetric measurements. Afterwards, electrodes were carefully washed in acetonitrile and a 10^{-4} mol dm^{-3} solution of acetic acid to remove the electrolyte residue.

Results and Discussion

Antireflective transparent coatings, and tunable or broadband absorption layers

Transparent conducting oxide NWs were grown using a unique self-seeding process during MBE deposition that results in defect-free branched NW layers whose composition can be tuned as well as the graded porosity and thus refractive index. For a given angle, the internal optical scattering (a balance of Rayleigh scattering, material absorption and effective medium graded index effects) allows a tunable absorption depending on the NW size, distribution, branching and overall photonic morphology. An example of the NW layer is shown below in Fig. 1a.

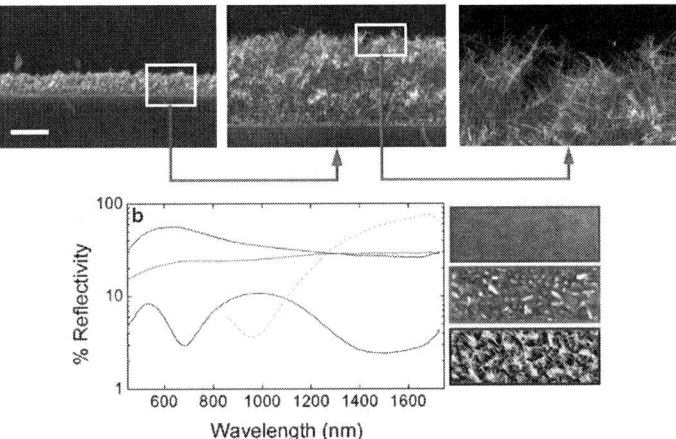

Figure 1. (a) A typical ITO NW layer grown on Si(100). The layer comprises a high density of branched NWs, which grow from a near compact contact layer to Si, followed by increased porosity from lower density growth away from the substrate. (b) Reflectance spectra for (———) a thin ITO antireflection coating, (————) a porous ITO layer, (········) an ITO NW layer (all on silicon) and (———) a thin ITO film on glass. Plan view SEM images of each layer are also shown.

In terms of evaluating the improvement afforded by such NW layers as advanced electro-optical device top contacts, the calculated reflectivity at near-normal incidence (10°) as a function of wavelength was determined and is shown in Fig. 1b for the ITO NW layer, a porous ITO layer, a thin ITO film and a standard antireflection layer of ITO on glass. ITO NW layers on silicon exhibited much lower reflectivity than either dense ITO on silicon or porous ITO layers in the wavelength region of interest. Only antireflection ITO coatings on glass showed similar transmission between 850-975 nm, and a usefully low sheet resistivity of ~15 Ω sq^{-1}. Fresnel reflection associated with higher refractive index ITO thin films is virtually eliminated (50) due to reduction in the refractive index from the substrate to the NW layer/ambient interface, where the refractive index was measured to vary from 2.19 (close to the substrate) to 1.04 (at the air interface). The low refractive index NW layer is optically specular and exhibits a 'dark' surface, confirming the absence of Fresnel reflection over a broad spectral width. This is particularly important for LED applications because of the isotropic emission from the active region and the fact that reflection phenomena can limit the light-extraction efficiency.

Typically, ITO and other transparent conducting oxides fail to transmit light in the near infrared region above ~1.1 μm, thus preventing their employment as electrodes for LEDs operating in the second (1.3 μm) and third (1.55 μm) optical communications energy windows. This is due to the onset of a metal-like behaviour at low photon energies, where quanta of incident electromagnetic radiation can couple to plasma oscillations. The surface plasmon resonance (SPR) of ITO (determined using the Drude free-electron model and three-phase Fresnel equations of reflection) is observed at a much lower energy than for metals (where it generally falls into the visible range). In Fig. 1b, the surface plasmon resonance of the ITO NW layer (6895 cm^{-1}) compared to a thin ITO film (10204 cm^{-1}) is red-shifted by ~480 nm, pushing the onset of reflectance further towards the near to mid-IR. The minimum in reflectivity is routinely observed in the crucial 1.2-1.6 μm range at near-normal incidence (0-10°). Angle-resolved reflectance measurements (50) conclusively show that the position of the surface plasmon resonance blue-shifts with increasing angle; the window for maximum transmission over the widest wavelength range is observed near-normal to the surface. Thus, the optical benefits of porosity afforded by the NW growth offsets the standard trade-off between high transparency and low conductivity (and vice versa) by extending conductive transparency into the infra-red, while allowing tunability in the absorption and angle dependent antireflection through structure.

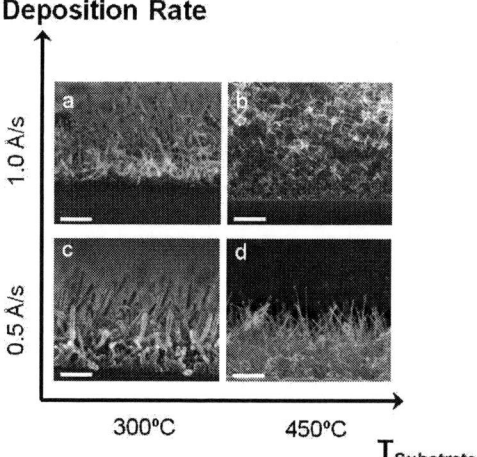

Figure 2. Cross-sectional SEM images of ITO NW layers deposited on silicon as a function of deposition rate and substrate temperature. For all structures, the In:Sn ratio (90:10) was evaporated at a temperature for In and Sn of T_{In} = 835°C, T_{Sn} = 1000°C, respectively in an oxygen pressure of 2.1 × 10^{-4} mbar. NWs range between 8-50 nm and are monocrystalline. Scale bars = 200 nm.

The growth rate and temperature are key parameters in the eventual morphology of the ITO NW layers, and these effects on morphology for a defined composition are summarised in Fig. 2, where consistently finer NWs are found at the higher growth rates and temperatures for a given composition, giving a gamut of similar structures with various porosities that allow tunable absorption or reflection across the visible range.

Indium oxide (In_2O_3) and SnO_2 were also grown as a range of nanodot dispersions. These dispersions have their effective optical constants altered by their effective porosity, characterized by their modal size distributions details of which be found elsewhere (51). An example of In_2O_3 and SnO_2 dispersions are shown in Fig. 3 below alongside ITO nanodots which were possible by halting the intial stages of NW seed formation.

In_2O_3, as a wide bandgap semiconductor ($E_g^d \sim 3.6$ eV; $E_g^i \sim 2.6$ eV), has a transparency that is fundamentally interlinked with it conductivity; the transition occurs on or around its plasma frequency related to the electron density according to the Drude model by $\omega_p^2 = nq^2/m^*\varepsilon_0$, where m^* and q are the effective mass and charge of an electron, and ε_0 the high-frequency permittivity of free-space. For the single-sided In_2O_3 nanodot dispersion, the transmission is limited by Fresnel reflection due to the step discontinuity in refractive index at the rear flat surface (52,53).

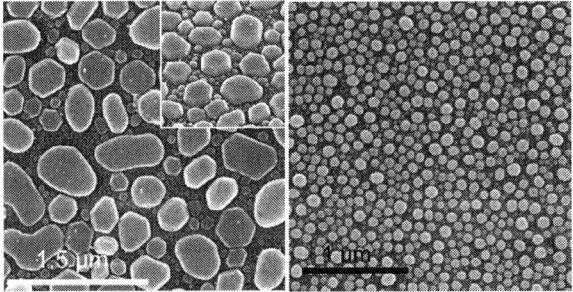

Figure 3. Morphology, distribution and shape differences between In_2O_3 (left) and ITO (right) nanodot dispersions. In_2O_3 are always faceted resulting from post-growth oxidative crystallization of hot substrate dewetted In metal liquid film grown by MBE. The ITO dots form in an oxygen atmosphere from co-deposition of In and Sn.

The In_2O_3 dots, unlike ITO and SnO_2 dots, have large crystals that have a size distribution covering the visible wavelength range, thus causing a transition to Mie scattering which results in the white or grey appearance, rather than a defined colour which varies with angle noted for other dots of smaller sizes and distribution of those sizes. Details on the size and modal distribution can be found elsewhere (51). These layers offer excellent visible to infra-red transmission, and also antireflection properties, as seen in the optical images in Fig. 4a.

Figure 4b shows the integrated total reflectance from a SnO_2 and In_2O_3 nanodots dispersion in the visible and infra-red regions, together with a reference ITO film on glass which exhibits significant increase in transparency as expected in the infra-red. The red-shifting of the entire angle-resolved spectrum for the In_2O_3 nanodot dispersions is also seen in Fig. 4c, where at visible wavelengths the transmission varies from 55% at near incidence, and importantly, remains 87% transparent at wavelengths up to 4 μm compared to the standard transmission of an ITO film on glass.

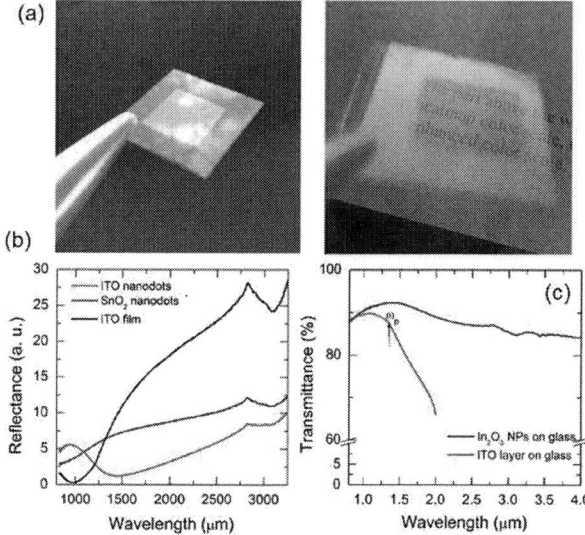

Figure 4. (a) Optical images of the angle-dependent antireflection properties of In_2O_3 nanodot dispersions on a glass cover slip. (b) Integrated total reflectance of ITO, SnO_2 and In_2O_3 nanodot dispersions in the infra-red region and (c) optical transmission characteristics of the In_2O_3 nanodots, all compared to a thin ITO film on glass.

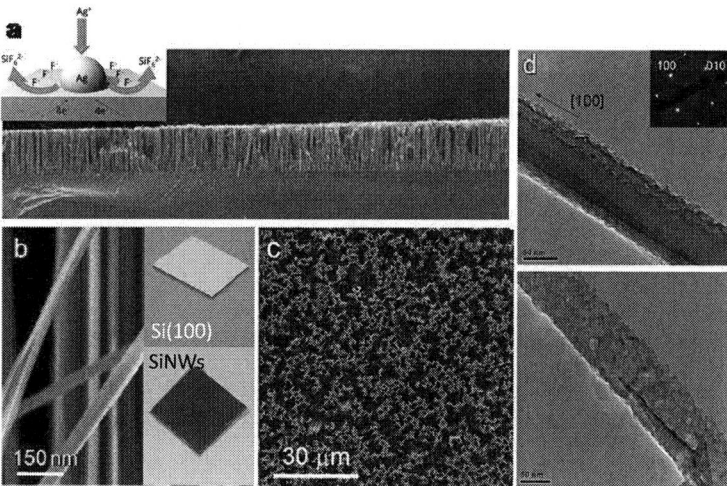

Figure 5. (a) (*Inset*) Schematic of the electroless metal deposition process. Cross-sectional SEM image of the Si NW at low magnification showing the uniformity of the etching depth and NW length. (b) Cross sectional SEM image of several NWs with inset photographs showing pristine Si(100) wafer and wafer with etched NWs. (c) Plan view SEM image of the NW layer. (d) TEM images of individual NWs showing the characteristic rough morphology of MAC etched NWs.

Silicon NW layers that offer good broadband absorption, and a near blackbody response at certain angles to white light, are possible by electrolessly etching single crystal silicon until a disordered array of rough NWs with a wide diameter distribution are formed. Figure 5 shows the resulting SiNW layers that are formed using the metal-assisted chemical (MAC) etching approach described above, and summarized schematically in Fig. 5a. Vertical arrays of NWs are formed across the entire surface of the wafer, as shown in the cross-sectional SEM image in Fig. 5a. High resolution SEM examination reveals that the NWs have an average length of 115 μm. Examination of the top surface of the Si NW layers shows that the wires appear in non-uniform distributions characterized by clumped regions of high density NWs and correspondingly, regions of locally less density (Fig. 5c). Further examination shows that the less dense regions are not devoid of NWs, rather the NWs are bent towards each other forming the high density regions. This stems from a combination of high length-to-width aspect ratio and capillary forces related to the post-etch cleaning procedure. NWs were measured to have an average width of 80 nm, shown in Fig. 5b. TEM analysis in Fig. 5d also confirmed that individual Si NWs are single crystal and electron diffraction measurements also confirm their (100) orientation, consistent with etching from bulk Si(100). The NWs have a characteristic rough morphology consistent with MAC etched NWs in previous reports. Figure 5b shows typical SEM and optical images (inset) of the pre and post-etched Si, where at angles between 45 and 55°, a completely black, strongly absorbing layer is visible. Such layers are being sought for silicon based photovoltaics, and their broadband absorption characteristics are typical of marginally controlled etching characteristics (facilitated by the reduction potential of Ag ions in HF solutions), giving a highly disorder array on several lengths scales (porosity of the semiconductor crystal, roughness on the walls of the wires, bent and clumped NWs randomly distributed throughout the vertical array etc., that facilitate a very high degree of optical scattering effects.

Bottom-up growth of fully transparent contact layers of indium tin oxide NWs for LEDs

One particular application of ITO that employs its conductive and transmissive properties, is that of an ohmic contact for LEDs, but has until now, only had limited success due to high resistivities and an unacceptable trade-off between its electrical and optical characteristics. The limited availability of materials with desired refractive indices, however, has prevented the implementation of optical components with very high performance. The possibility of forming controlled complex, multi-level branched structures with optimized optical properties in a single-step, bottom-up growth regime, is a real advance considering their compatibility with optoelectronic device architectures.

We have developed a controllable molecular beam epitaxial growth system for obtaining high quality, uniphasic, branched ITO NWs on silicon and oxidised silicon surfaces using In and Sn precursors in an oxygen atmosphere. The NW layer morphology and transparency can be tailored through rational control of the evaporation-condensation parameters, and will be shown to result in bottom-up grown layers of self-catalysed and seeded NWs with excellent electrical, optical and homogeneous structural properties, that can even be grown as large area (several cm²) contacts. The graded porosity NW layers are grown as an alternative option for the standard NiCr top contact on a Si/SiGe multi-quantum well (MQW) LED, exhibiting a transparency of >80 % in the visible and >90% in the 1.2–1.6 μm range.

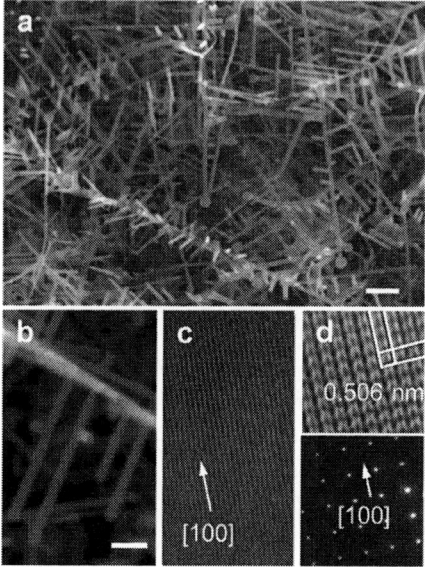

Figure 6. (a) FESEM image of the interior of a typical ITO NW array grown at T_{sub} = 500 °C at nominal In:Sn (90:10) growth rates of 0.1 nm s^{-1} highlighting the mutually orthogonal [100] branching from the principal ITO NW backbone in three dimensions. Scale bar = 50 nm. (b) Higher magnification images of homogenous branching events. The scale bars = 25 nm. (c) HRTEM image of an individual NW highlighting interplanar spacings and the [100] growth direction. (d) The higher magnification HRTEM image and associated SAED pattern identify monocrystalline $(In_{1.875}Sn_{0.125})O_3$ with (002) and (020) interplanar spacings of 0.506 nm.

Various morphologies for graded refractive index coatings are possible by choosing any of the NW morphologies or nanodot dispersions according to deposition rate and substrate temperatures. For NWs, a distinct growth mode modification from compact (but rough and partially porous) layers to dendritic NW growth during deposition is achievable at specific growth rates and substrate temperatures. Locally increased porosity is the main defect found in graded index porous layer growth by oblique angle deposition techniques; such problems are not encountered for dense, highly branched NW layers as outlined below in Fig. 6. No heterogeneous catalysts are required and we still observe extremely high quality, defect-free NWs comprising the layers.

After the growth was optimized on both quartz and silicon substrates, the growth method was applied to corresponding n–i–p$^+$ Si/SiGe MQW device structures, which in the device employed in this work consisted of an active region with a 15 period Si/Si$_{0.77}$Ge$_{0.23}$/Si MQW layer. An example of an ITO NW layer grown on the Si/SiGe MQW structure is shown in the SEM image in Fig. 7a with a corresponding device schematic shown in Fig. 7b.

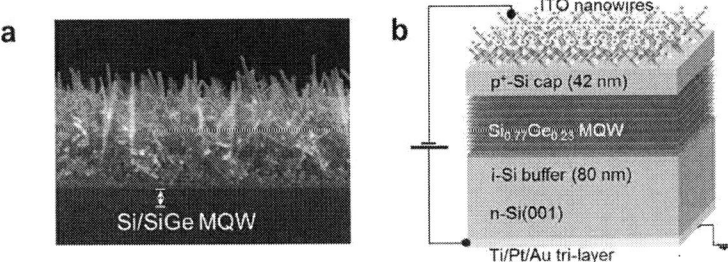

Figure 7. (a) Cross-sectional SEM images of the NW layer on a SiGe MQW. The NW layer shown was grown at a substrate temperature of $T_{sub} = 575$ °C at a growth rate of 0.2 nm s^{-1} and an invariant oxygen partial pressure of 2.1 × 10^{-4} mbar. (b) Schematic representation of the ITO NW contacted SiGe MWQ LED. The active region consists of a 15 period Si/Si$_{0.77}$Ge$_{0.23}$/Si multiple quantum well layer. The bottom contact is a Ti/Pt/Au trilayer. The NWs were grown directly onto the p$^+$-Si.

To assess this influence on the optical transmission, ITO layers of varied thickness and morphology were grown on quartz substrates. Figure 8a shows their transmission spectra in the wavelength range 400-1600 nm. The inset shows a typical ITO NW layer deposited on a glass cover slip. The transparency in the visible range improves to ~90% with increasing growth rate up to 0.2 nm s^{-1}. Spectroscopic ellipsometric measurements show effective refractive indices of the NW layer-air interface to be between 1.04–1.12 for all samples investigated, directly due to the varied porosity of the nanostructured layers, calculated using the Bruggemann effective medium approximation. This is very advantageous as previous efforts to couple high transmission and low resistivity resorted to ITO/Ag/ITO multilayer sandwiches to combine the improved conductivity of Ag and the high refractive index of the ITO ($\eta = 2.19$) to boost transmittance of the metal interlayer (54,55). The measured absorption edge for the nanostructured ITO films is ~380 nm, a value lower than that for ITO on glass and comparable to ITO/YSZ lattice matched thin films and NWs (56).

Their suitability to silicon-based LEDs, is demonstrated by investigation of the room temperature electroluminescence (EL) emission from the NW contacted SiGe MQW LED, shown in Fig. 8a, showing unambiguously that the Si and SiGe emission lines are fully resolved through the NW layer contact in ambient laboratory conditions. The corresponding comparative spectra of both the uncoated MQW and nanostructured ITO-contacted MQW are shown in Fig. 9a, where a marked improvement in transmission compared to the NiCr contacted MQW is observed for electroluminescence emission. The highest attenuation was noted for the Si^{EHD} emission, although 20% of the uncoated MQW signal for this line was still transmitted, as opposed to negligible transmission through the NiCr contact. The Si$_{BE}$^{TO} emission was attenuated by ~50% but of utmost interest are the SiGe emission lines, which were transmitted with negligible absorption. It is clear that the full spectrum (all important transitions) of a SiGe MQW LED can be resolved through the ITO NW top contact.

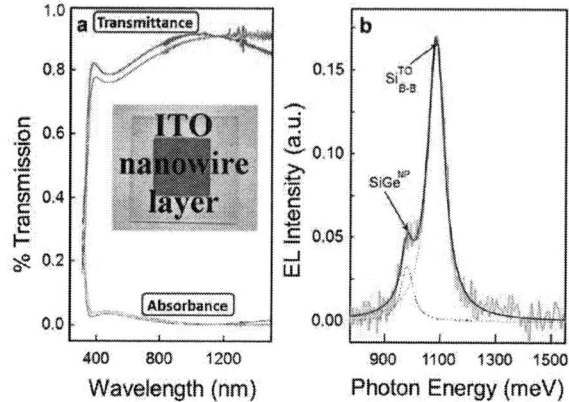

Figure 8. (a) Transmission measurements of an ITO NW layer deposited at a substrate temperatures of T_{sub} = 575 °C (———) and T_{sub} = 450 °C (———) at a nominal growth rate of 0.05 nm s^{-1} and an invariant oxygen partial pressure of 2.1 × 10^{-4} mbar. The inset shows an ITO NW layer deposited on a glass cover slip. (b) Room temperature (295 K) electroluminescence (EL) spectrum from an ITO NW-contacted Si/SiGe MQW structure acquired at an injection current of 575 mA. The SiGe^{NP} (no phonon) and Si$^{TO}_{B-B}$ (band-to-band transverse optical phonon) emission were extracted with a multiple Gaussian fit.

The geometry of the NW layer maximizes transmission through a single-step, practical approach of randomizing scattering photons. In addition, the emission is maximised by the angular dependence of the plasma frequency, where the transmission is both maximized and limited to near-normal output. The optical and electrical properties as a whole are unprecedented for any conductive, transparent oxide and have the added benefit of being experimentally much simpler than more complex, multi-step deposition of several defect-free layers of reducing refractive index (exhibiting poor conductivities) and highly conducting thin films (exhibiting poor transmission in the visible region).

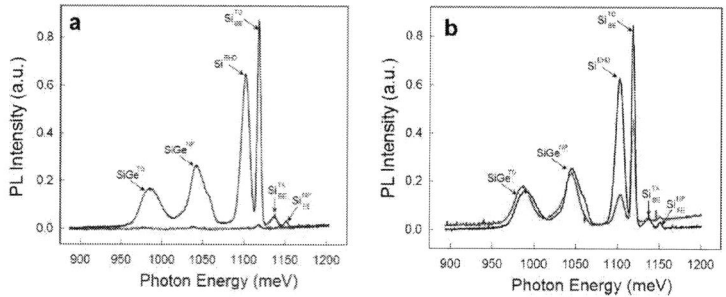

Figure 9. Comparison of the low temperature (45 K) photoluminescence spectra of (a) a non-contacted SiGe MQW (———) and a NiCr contacted SiGe MQW (———) (b) a non-contacted SiGe MQW (———) and an ITO NW-contacted SiGe MQW (———). The two high intensity peaks observed at 1125 and 1150 nm correspond to the silicon transverse optical phonon-assisted bound exciton recombination (Si$_{BE}$ ^{TO}) and the silicon electron-hole droplet emission (Si^{EHD}), respectively.

Phonon engineering in Si nanostructures for thermoelectrics

It has been shown that when crystalline solids are confined to the nanometer range, electron and phonon transport can be significantly altered due to three discrete effects (57) such as increased boundary scattering, changes in phonon dispersion, and quantization of phonon transport. Similarly, theoretical and experimental work (58) has shown that the electrical and thermal conductivities from NWs differ from those for bulk Si (59). Furthermore, transport properties are very sensitive to the crystalline lattice characteristics of individual NWs. Consequently, systematic structural characterization is necessary to distinguish these characteristics.

Silicon has not been considered a viable thermoelectric energy converter due to its exceptionally high thermal conductivity. Heat carried by phonons in Si have large mean-free-path lengths >200-300 nm (60). Reducing the dimensionality of all or part of a crystal has been shown to be an effective method for increasing thermal resistivity (61-65). For Si at room temperature, useful thermoelectric figures of merit need a lattice thermal conductivity of ~1W/m-K. Thermal conduction limited by grain boundary scattering processes predict that small feature sizes (~2 nm) can make Si a viable thermoelectric (66,67), and methods to engineer such sizes are a focus of research interest.

Using MAC etched Si NWs described in Fig. 5, nanostructuring of Si NWs on more than one length scale allows for a significant increase in thermal resistance, beyond Umklapp process and boundary scattering. By etching Si with a defined resistivity, the resulting NWs can exhibit a high degree of roughness defined by small crystallites of silicon that are not simple edges of rough features delineated by the anisotropic electroless etching process (68-70). Lightly doped p-type Si results in rough NWs coated with a high density of crystallites that are oriented in a range of zone axes different to that of the NW itself, as shown in the bright field and high angle annular dark field image of a rough Si NW in Fig. 10a and b. These structures are effective phonon confining crystallites that also increase the boundary scattering for phonons, thereby increasing the thermal resistance.

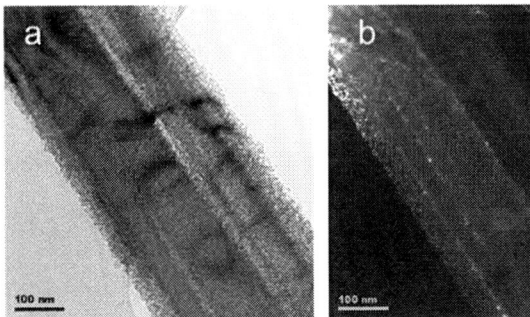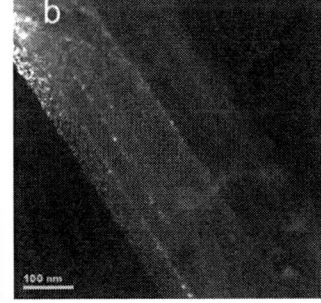

Figure 10. (a) Bright field and (b) high angle annular dark field TEM image of a rough p-type Si NW.

For horizontally oriented Si NWs, a deconvoluted contribution to the asymmetry in the TO phonon peak is found at 506 cm^{-1} (Figure 11a). This deconvoluted band at 506 cm^{-1} is not expected for nanosized diamond cubic Si. It has been shown that this band is due to the presence of hexagonal Si structure. This agrees well with the Raman measurements made on hexagonal diamond Si obtained by nanoindentation (71). The results are also comparable to the measurements realized in microcrystalline samples with a mixture of cubic and hexagonal diamond Si phases in agreement with semi empirical models. In our case, contributions at these frequencies due to oxide presence cannot explain the observation of the convoluted band at elevated temperature, despite evidence that a non-uniform oxide is present along the length of all wires investigated.

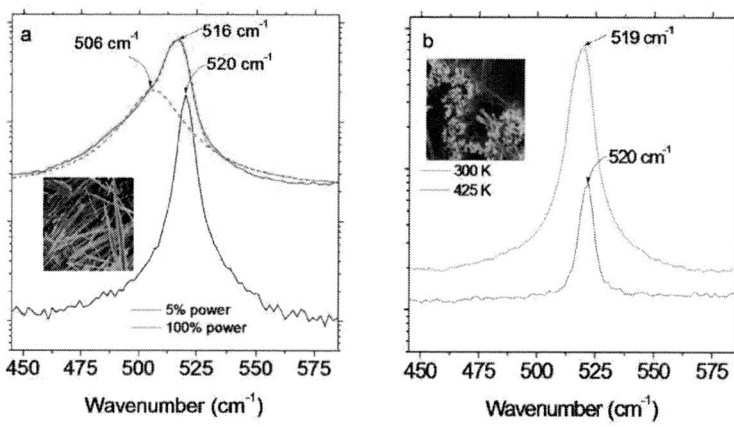

Figure 11. (a) Raman scattering of the TO phonon from (a) v-SiNWs and (b) h-SiNWs as a function of temperature. Spectra were acquired at 5, 10 and 100% of incident laser power.

In Fig. 12, the TO mode is shown for four of the spectra, as marked in Fig. 12a. We observe the characteristic small redshift from 521 to ~519 cm^{-1} which reverts to 521 cm^{-1} when the spectrum is again taken from the vertically oriented Si NWs. This confirms the contribution from confining crystallites along the outer surface of the NWs; their contribution is seen when their inelastic scattering cross section is increased during probing of the sides of the NWs.

Future work is ongoing to define the thermal resistance changes in Si NWs with increased surface roughness that enhances the red-shifting and asymmetric broadening of their Raman spectra. SiO_x contributions were not evident and specific substrate Raman modes were suppressed for horizontal NWs. The correlation length L estimated by this model is related to the average grain size in nanocrystalline materials or to the average distance between defects in crystals but alone does not give a satisfactory description of the Raman band-shape modification for the data. However, the ability to strongly affect phonon transport results in inelastic scattering characteristics that are directly related to a significant lowering of the thermal conductance augers well for optimization of simple and quick electrochemical processes for engineering of phonon transport in Si nanostructures for thermoelectrics.

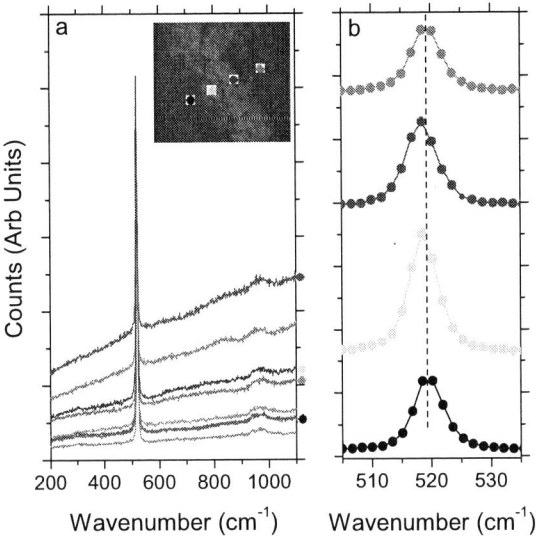

Figure 12. (a) Raman scattering spectra acquired by crossing from regions of v-SiNWs to h-SiNWs and then to v-SiNWs. (*Inset*) Optical image taken from the spectrometer showing the h-SiNWs (light region) and vertical NWs (dark flanking regions) and points of Raman measurements. (b) Variation in TO phonon scattering frequency as a function of position shown in the optical image in (a).

Si, In_2O_3 and SnO_2 nanostructures as anode materials for Li-ion batteries

To increase the energy content of lithium-ion batteries, significant research has been devoted to finding higher capacity electrode materials. On the cathode side, sulphur and oxygen-based positive electrodes have recently been intensely researched (72-74). On the anode side, alloy anodes have shown significant promise. An ideal anode material should possess high gravimetric and volumetric capacity, a low potential against cathode materials, long cycle life, light weight, environmental compatibility, low toxicity, and low cost (75). Silicon has performance that make it a very attractive choice as an anode material, given its ability to mitigate volumetric expansion pulverization that can result in electrical disconnect and disintegration during operation over many cycles, and the fact that it has the highest theoretical specific capacity for Li^+ intercalation.

We have studied Si NW layers by comparison to associated bulk Si(100) samples to deduce the direct effect of nanostructuring on identically doped Si where all NWs formally have the same crystal orientation. In addition, the ability of porous and graded index transparent conducting oxides such as In_2O_3 and SnO_2 similar to those shown in Fig. 3, which are excellent alloy materials, represent some of the most promising anode reactions for safe high capacity Li-ion batteries. These dispersions also allow for the possibility of optical probing of intercalation and related electrochemical processes. Both transparent oxide materials allow rechargeable alloying with Li, resulting in stable and efficient reversible charge storage. For both materials, the dispersion of sizes is critical in their evaluation and analysis towards future development.

Figure 13 shows the first five voltammetric cycles of an SnO_2 nanodot electrode, and we note that the reduction of SnO_2 to Sn^0, allows an alloy reaction with Li. The cyclic voltammograms of SnO_2 show that the cathodic process involved the insertion of Li into Sn_2O_3 to form a Li–Sn alloy (charging) and the anodic process follows Li extraction or dealloying (discharging). Details of the In_2O_3 electrode can be found elsewhere (76). The first cycle for each of the dispersions confirms SEI layer formation at the higher cathodic potentials. For both materials, the Li insertion potentials remain quite low. A related process is known to occur for Li alloying with Sn (77).

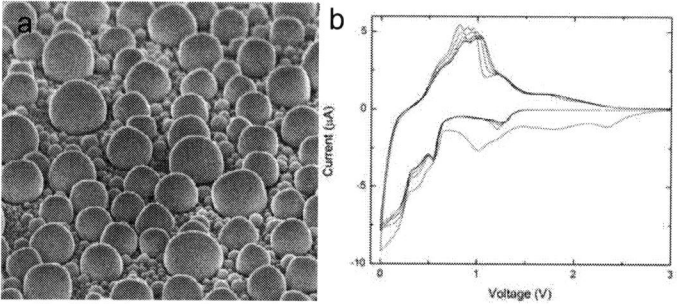

Figure 13. SEM image of epitaxial non-faceted SnO_2 nanodots on Si(100) with corresponding cyclic voltammograms.

Here, the dispersion of the nanodots, while permitting transparency and conductivity into the IR region, also allows lithium insertion into the current collector, and remains structurally stable due to strong epitaxial adhesion and lithium co-insertion buffering by the silicon current collector. The partial porosity of the variable size nanodot distribution prevents stress build-up in the In^0 and Sn^0 (when reduced from In_2O_3 and SnO_2) and promotes optical transparency. The removal of the smallest nanodots by volume expansion of the underlying silicon does not greatly affect the gravimetric energy density, and it serves to buffer polarization of the Li_xIn and Li_xSn by reversibly inserting Li^+.

Si NWs also offer the possibility of reversible Li insertion, even when etched from Si wafers. Figure 14 shows cyclic voltammograms of a Si(100) electrode and a Si NW electrode etched from the same structure, also shown in Fig. 14. Si NWs on a silicon current collector exhibit reversible lithium insertion potentials, identical to those from the bulk Si(100) substrate. The Si(100) undergoes some surface morphological changes, where cracking of the surface occurs, freeing up fresh Si for intercalation, resulting in self-similar cyclic voltammetric response with a good coulometric efficiency. For lower doped Si(100), however, we note that cracking is much less evident, due to a reduced Li^+ uptake by a lower carrier (electron) density in the p-type materials compared to n-type. For the NWs, considerable space charge capacitance contributed to the measured charge, and further work is underway to define the transition from intercalation to pseudocapacitance for these materials.

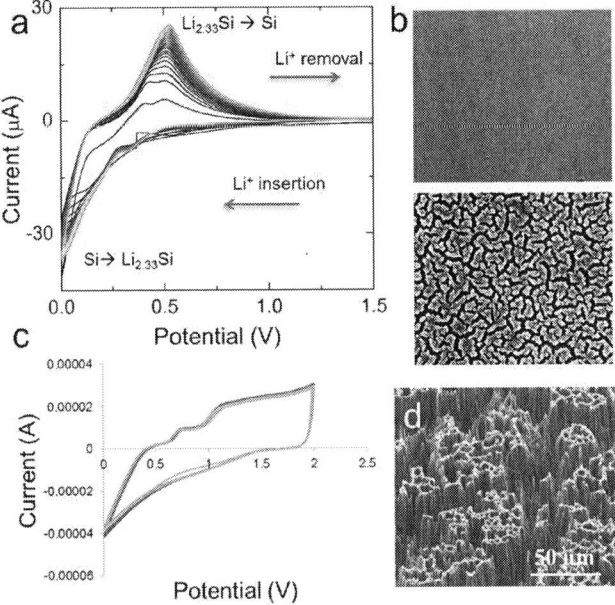

Figure 14. (a) Cyclic voltammograms of p-type Si(100) in a 1 mol dm^{-3} electrolyte containing LiPF$_6$ in EC:DMC (50:50 v/v) at a scan rate of 1 mV s^{-1}. (b) SEM image of the p-type (top) and n-type (bottom) surfaces following 30 potentiodynamic cycles. (c) Cyclic voltammograms of the p-type Si NWs, and (d) SEM image of the p-type Si NWs.

Conclusions

The fabrication of high aspect ratio monocrystalline and highly transparent doped metal-oxide NW layers and electroless etching of silicon into vertical NW layers with unprecedented simplicity in morphological control and phase homogeneity was presented. The NWs prove to be excellent materials for broadband absorption, tunable reflection characteristics useful for optoelectronic and related devices, but also have useful structural and electrochemical properties that allow reversible charge storage as Li-ion battery anodes. The roughness of Si NWs also has a marked effect on phonon transport, allowing for lower thermal conductance and the possibility for junctionless power from temperature differences. Structuring semiconductors through top down electrochemical means or through grading porosity through bottom up growth methods, allows for a variety of useful electrochemical electrical, optical and structural effects that are amenable for further consideration in energy related materials research.

Acknowledgements

This work was supported by Science Foundation Ireland (SFI) under contract No. 07/SK/B1232a. CG, MO and EA acknowledge financial support from the Irish Research council under Awards No. RS/2011/797, RS/2010/2170, and RS/2010/2920. EA acknowledges SFI for a Short Term Travel Fellowship under award No.

07/SK/B1232a-STTF11. Part of this work was conducted under the framework of the INSPIRE programme, funded by the Irish Government's Programme for Research in Third Level Institutions, Cycle 4, National Development Plan 2007-2013. The authors also acknowledge financial support from the European Union 7th Framework Programme under the SiNAPS project (project ref: 257856).

References

[1] H. Föll, S. Langa, J. Carstensen, M. Christophersen and I.M. Tiginyanu, *Adv. Mater.* **15**, 183 (2003).

[2] E. Spiecker, M. Rudel, W. Jäger, M. Leisner and H. Föll, *phys. stat. sol. (a)* **202**, 2950 (2005).

[3] S. Langa, J. Carstensen, I.M. Tiginyanu, M. Christophersen and H. Föll. *Electrochem. Solid-State Lett.* **5**, C14 (2002).

[4] M. Christophersen, S. Langa, J. Carstensen, I.M. Tiginyanu and H. Föll, *phys. stat. sol. (a)* **197**, 197 (2003).

[5] S. Lölkes, M. Christophersen, S. Langa, J. Carstensen and H. Föll, *Mater. Sci. Eng. B*, **101**, 159 (2003).

[6] S. Langa, J. Carstensen, M. Christophersen, K. Steen, S. Frey, I.M. Tiginyanu and H. Föll, *J. Electrochem. Soc.* **152**, C525 (2005).

[7] F.M. Ross, G. Oskam, P.C. Searson, J.M. Macaulay and J.A. Liddle, *Philos. Mag. A* **75**, 525 (1997).

[8] H. Tsuchiya, M. Hueppe, T. Djenizian and P. Schmuki, *Surf. Sci.* **547**, 268 (2003)

[9] H. Tsuchiya, M. Hueppe, T. Djenizian, P. Schmuki and S. Fujimoto, *Sci. Tech. Adv. Mater.* **5**, 119 (2004).

[10] S. Langa, I.M. Tiginyanu, J. Carstensen, M. Christophersen and H. Föll, *Appl. Phys. Lett.* **82**, 278 (2003).

[11] S. Langa, M. Christophersen, J. Carstensen, I.M. Tiginyanu and H. Föll, *phys. stat. sol. (a)* **197**, 77 (2003).

[12] A.-M. Gonçalves, L. Santinacci, A. Eb, C. David, C. Mathieu, M. Herlem and A. Etcheberry, *phys. stat. sol. (a)* **204**, 1286 (2007).

[13] A.-M. Gonçalves, L. Santinacci, A. Eb, I. Gerard, C. Mathieu and A. Etcheberry, *Electrochem. Solid-State Lett.* **10**, D35 (2007).

[14] M.M. Faktor, D.G. Fiddyment and M.R. Taylor, *J. Electrochem. Soc.*, **122**, 1566 (1975).

[15] E. Harvey, C. Heffernan, and D. N. Buckley and C. O'Raifeartaigh, *Appl. Phys Lett.*, **81**, 3191 (2002).

[16] L. Macht, J.J. Kelly, J.L. Weyher, A. Grzegorczyk and P.K. Larsen, *J. Crystal Growth* **273**, 347 (2005).

[17] J. J. Kelly, L. Macht, D.H. van Dorp, M.R. Kooijman and J.L. Weyher, *in Processes at the Compound-Semiconductor/Solution Interface*, Editors: P.C. Chang, K. Shiojima, R.E. Kopf, X. Chen, D. Noel Buckley, A. Etcheberry, and B. Marsan, State-of-the-Art Program on Compound Semiconductors XLII, **PV 2005-04**, p. 138 (2005).

[18] R. Lynch, C. O'Dwyer, D.N. Buckley, D. Sutton and S.B. Newcomb, *ECS Trans* **2**, 131 (2006).

[19] R. Lynch, M. Dornhege, P. Sánchez Bodega, H.H. Rotermund and D.N. Buckley, *ECS Trans* **6**, 331 (2007).

[20] R. Lynch, C. O'Dwyer, D. Sutton, S. Newcomb, and D.N. Buckley, *ECS Trans* **6**, 355 (2007).

[21] R. Lynch, C. O'Dwyer, N. Quill, S. Nakahara, S.B. Newcomb, and D.N. Buckley, *ECS Trans* **16**, 393 (2008).

[22] N. Quill, C. O'Dwyer, R. Lynch, C. Heffernan, and D. N. Buckley, *ECS Trans* **19**, 295 (2009).

[23] C. O'Dwyer, D.N. Buckley, D. Sutton, and S.B. Newcomb, *J. Electrochem. Soc.,* **153**, G1039 (2006).

[24] C. O'Dwyer, D.N. Buckley, and S. B. Newcomb, *Langmuir*, **21**, 8090 (2005)

[25] C. O'Dwyer, D.N. Buckley, D. Sutton, M. Serantoni, and S.B. Newcomb, *J. Electrochem. Soc.,* **154**, H78 (2007).

[26] O. V. Bilousov, J. J. Carvajal, D. Drouin, X. Mateos, F. Díaz, M. Aguiló and C. O'Dwyer, *ACS Appl. Mater. Interfaces*, **4**, 6927 (2012).

[27] C. K. Chan, H. L. Peng, G. Liu, K. McIlwrath, X. F. Zhang, R. A. Huggins, Y. Cui, *Nat. Nanotechnol.* **3** (2008) 31 (2008).

[28] M. T. McDowell, S.W. Lee, C. Wang, Y. Cui, *Nano Energy* **1**, 401 (2012)

[29] M. T. McDowell, Y. Cui, *Adv. Energy Mater.* **1**, 894 (2011).

[30] J. Y. Tang, H. T. Wang, D. H. Lee, M. Fardy, Z. Y. Huo, T. P. Russell, and P. D. Yang, *Nano Lett.* **10**, 4279 (2010).

[31] H. S. Yang, D. G. Cahill, X. Liu, J. L. Feldman, R. S. Crandall, B. A. Sperling, and J. R. Abelson, *Phys. Rev. B* **81**, 104203 (2010).

[32] A. I. Hochbaum, R. K. Chen, R. D. Delgado, W. J. Liang, E. C. Garnett, M. Najarian, A. Majumdar, and P. D. Yang, *Nature* **451**, 163 (2008).

[33] D. S. Ginely, C. Bright, *MRS Bull.* **25**, 15 (2000).

[34] Y. Cui, C. M. Lieber, *Science* **291**, 851 (2006).

[35] Y.-J. Hsu, S.-Y. Lu, *J. Phys. Chem. B* **109**, 4398 (2005).

[36] J. G. Lu, P. Chang, Z. Fan, *Mater. Sci. Eng. R* **52**, 49 (2006).

[37] M. C. Johnson, S. Aloni, D. E. McCready, E. Bourret-Courchesne, *Cryst. Growth Des.* **6**, 1936 (2006).

[38] P. Nguyen, et al. *Nano Lett.* **3**, 925 (2003).

[39] Q. Wan, et al. *Nano Lett.* **6**, 2909 (2006).

[40] D.-W. Kim, et al. *Nano Lett.* **7**, 3041 (2007).

[41] D. L. Wang, C. M. Lieber, *Nature Mater.* **2**, 355 (2003).

[42] J.M. Tarascon, M. Armand, *Nature* **414**, 359 (2001).

[43] M.S. Whittingham, *Chem. Rev.* **104**, 4271 (2004).

[44] M. Armand, J.M. Tarascon, *Nature* **451**, 652 (2008).

[45] D. Linden, T. Reddy, Handbook of Batteries, McGraw-Hill, New York, 2002.

[46] J. B. Goodenough, H. Abruna, M. Buchanan, US Department of Energy, Washington, DC, 2007.

[47] C. Liu, F. Li, L.P. Ma, H.M. Cheng, *Adv. Mater.* **22**, E28 (2010).

[48] Y. J. Lee, H. Yi, W.J. Kim, K. Kang, D. S. Yun, M. S. Strano, G. Ceder, A. M. Belcher, *Science* **324**, 1051 (2009).

[49] B. Scrosati, J. Garche, *J. Power Sources* **195**, 2419 (2010).

[50] C. O'Dwyer, M. Szachowicz, G. V. Visimberga, V. Lavayen, S. B. Newcomb, C. M. Sotomayor Torres, *Nat. Nanotech.* **4**, 239 (2009).

[51] M. Osiak, W. Khunsin, E. Armstrong, T. Kennedy, C. M. Sotomayor Torres, K. M. Ryan, and C. O'Dwyer, *Nanotechnology*, **24**, 065401 (2013).

[52] J. Q. Xi, J. K. Kim and E. F. Schubert, *Nano Lett.* **5**, 1385 (2005).

[53] J.-Q. Xi, M. F. Shubert, J. K. Kim, E. F. Shubert, M. Chen, S.-Y. Lin, W. Liu and J. A. Smart, *Nature Photon.* **1**, 176 (2007).

[54] C. G. Granqvist, A. Hultåker, *Thin Solid Films* **411**, 1 (2002)

[55] S. Ju, *et al. Nat. Nanotech.* **2**, 378 (2007).

[56] Q. Wan, P. Feng, T. H. Wang, *Appl. Phys. Lett.* **89**, 123102 (2006).

[57] M. S. Dresselhaus, G. Dresselhaus, and A. Jorio, *Annu. Rev. Mater. Res.*, **34**, 247 (2004).

[58] D. Li, Y.Wu, P. Kim, L. Shi, P. Yang, and A. Majumdar, *Appl. Phys. Lett.* **83**, 2934 (2003).

[59] A. I. Hochbaum, D. Gargas, Y. J. Hwang, and P. Yang, *Nano Lett.,* **9**, 3350 (2009).
[60] K. Esfarjani, G. Chen, and H. T. Stokes, *Phys. Rev. B* **84**, 085204 (2011).
[61] L. Weber and E. Gmelin, *Appl. Phys. A* **53**, 136 (1991).
[62] D. Y. Li, Y. Y. Wu, P. Kim, L. Shi, P. D. Yang, and A. Majumdar, *Appl. Phys. Lett.* **83**, 2934 (2003).
[63] Y. S. Ju and K. E. Goodson, *Appl. Phys. Lett.* **74**, 3005 (1999).
[64] W. J. Liu and M. Asheghi, ASME Trans. *J. Heat Transfer* **128**, 75 (2006).
[65] P. E. Hopkins, C. M. Reinke, M. F. Su, R. H. Olsson, E. A. Shaner, Z. C. Leseman, J. R. Serrano, L. M. Phinney, and I. El-Kady, *Nano Lett.* **11**, 107 (2010).
[66] A. K. McCurdy, H. J. Maris, and C. Elbaum, *Phys. Rev. B* **2**, 4077 (1970).
[67] H. B. G. Casimir, *Physica* **5**, 495 (1938).
[68] O. Lotty, N. Petkov, Y. M. Georgiev and J. D. Holmes, *Jpn. J. Appl. Phys.* **51**, 11PE03 (2012).
[69] W. McSweeney, O. Lotty, J. D. Holmes, C. O'Dwyer, *ECS Trans.* **35**, 25 (2011).
[70] C. Glynn, O. Lotty, W. McSweeney, J. D. Holmes, C. O'Dwyer, *ECS Trans.* **8**, 73 (2011).
[71] A. Fontcuberta i Morral, J. Arbiol, J. D. Prades, A. Cirera, and J. R. Morante, *Adv. Mater.* **19**, 1347 (2007).
[72] Y. Yang, M.T. McDowell, A. Jackson, J.J. Cha, S.S. Hong, Y. Cui, *Nano Lett.* **10**, 1486 (2010).
[73] B. L. Ellis, K. T. Lee, L. F. Nazar, *Chem. Mater.* **22**, 691 (2010).
[74] X. L. Ji, K. T. Lee, L. F. Nazar, *Nat. Mater.* **8**, 500 (2009).
[75] J. R. Szczech, S. Jin, *Energy Environ. Sci.* **4**, 56 (2001).
[76] M. Osiak, W. Khunsin, E. Armstrong, T. Kennedy, C. M. Sotomayor Torres, K. M. Ryan, and C. O'Dwyer, *ECS Trans.* (*ibid.*) (2013).
[77] J. S. Chen, X. W. Lou, *Mater. Today* **15**, 246 (2012).

Characteristics of SnSbSe (SSS) Thin Films Grown by Atomic Layer Deposition for High Performance Phase Change Random Access Memory (PCRAM)

Keun Lee, Sehun Kang, Jachun Ku, Kwon Hong, and Sungki Park

NM Material Research Team, R&D Division, SK Hynix Semiconductor Inc., San136-1, Ami-ri, Bubal-eub, Icheon-si, Gyeonggi-do, 467-701, Korea

Characteristics of SSS(SnSbSe) thin films grown by atomic layer deposition(ALD) were investigated for high performance phase change random access memory(PCRAM). Adopting super-cycle concept which consist of SbSe binary and Sn single sub-cycle and adjusting Sb/Se exposure ratio, stoichiometry was successfully controlled. Microstructure and crystallization speed of ALD Sn-doped Sb-rich SbSe films were confirmed. Furthermore, ALD SSS film showed excellent gap-filling performance using confined cell structure of aspect ratio 5.0 (cell dimension 15nm).

Introduction

Phase change random access memory (PCRAM) has attracted great interest as a candidate for next generation non-volatile devices with the meet of increasing need for high density, low power consumption, and fast switching speed with CMOS logic process.[1,2] Reversible phase change between the amorphous (high resistance, RESET state) and crystalline (low resistance, SET state) of phase change material is basic operation principle in PCRAM. Ternary compound of Ge, Sb, and Te, especially $Ge_2Sb_2Te_5$ (GST225) has been widely investigated and accepted as a phase change material for the realization of PCRAM device. However, relatively slow crystallization (hundreds of nano-seconds) and low resistivity (hundreds of micro-ohm.cm^3) of GST limit the operation speed and writing current of PCRAM device.

As a storage element for high performance device, fast crystallization and high resistivity properties are required compared to typical GST material. In this point of view, many phase change materials (PCMs) have been widely investigated. Doping and element substitution are one of the effective PCM modification methods to reduce crystallization speed and increase resistivity. There have been several reports on the improvements of PCM properties by various doping or element substitutions, such as N[4-7], O[8], Si[9], SiO$_2$[10], In[11], Sn[12], Bi[13], Se[14], etc. In recent years, Se-based chalcogenide[15] material and Sn-doped PCM[12] have been studied for high-performance PCRAM device.

Furthermore, confined cell structure is adopted for high density and low power PCRAM device.[16] In this case, atomic layer deposition (ALD) process should be used to deposit the phase change material instead of the conventional chemical vapor deposition (CVD) process because high conformal growth needs to fill the narrow and deep hole. In addition, uniform doping within the hole is critical to the reliable operation and minimized cell to cell variation.

In this study, we investigated the growth behavior and microstructure of Sn-doped Sb-rich SbSe(SnSbSe) phase change materials deposited by thermal atomic vapor deposition (ALD) adopting a super-cycle concept.

Experiment

SSS(SnSbSe) thin films were deposited on Si_3N_4(40nm)/Si and bare Si substrate by thermal atomic layer deposition (ALD) using metal-organic precursors. We adopted a super-cycle concept to control stoichiometry and growth rate in multi-component thin films. One super-cycle for SSS consists of binary SbSe and single Sn sub-cycles. During the SbSe and Sn sub-cycle, cycles were repeated until the desired thickness was obtained and super-cycle can be represented as (SbSe:Sn) = (a:b), where a and b is the number of SbSe and Sn sub-cycles. SbSe and Sn layer were grown sequentially during the one super-cycle and their growth rate per cycle was controlled to fabricate the nano-mixed structure of SSS thin films. Various film compositions were obtained by controlling the sub-cycle ratio [$a/(a+b)$ or $b/(a+b)$] and the exposure ratio of precursors in each sub-cycle.

SnSb sub-cycle				Sn sub-cycle	
a-cycle				b-cycle	
Sb					
	Prg				
		Se			
			Prg		
				Sn	
					Prg

Figure 1. ALD SSS pulse sequence consist of [SbSe]-[Sn] super-cycle. Composition was controlled by ratio of a-cycle and b-cycle. [$a/(a+b)$ or $b/(a+b)$]

The cross section image and surface morphology of the film was investigated by a field-emission scanning electron microscopy (FESEM) and film roughness was confirmed by atomic force microscopy (AFM). The microstructure of the as-deposited film was investigated by transmission electron microscopy (TEM) and X-ray diffraction (XRD). Also, the crystallization behavior in nano-second scale was observed by using laser static tester.

Step-coverage and gap-filling of SSS films were confirmed using confined cell structure which has critical dimension of 15nm. (Aspect ratio 5.0:1)

Result and Discussion

As mentioned before, one super-cycle for Sb-doped Sb-rich SbSe consists of binary SbSe and single Sn sub-cycles. In order to deposit nano-mixed film and control Sn doping concentration, single Sn sub-cycle which consists of precursor feeding and purge step was separated. Several growth mode are possible in ALD process.[17] In *two-dimensional growth* (layer-by-layer growth, Frank-van der Merwe growth), one monolayer of the ALD-grown material covers the substrate completely. However, this growth mode is close to ideal case because full coverage could not be obtained considering metal-liand size on the substrate so called 'steric hindrance'. On the other hand, in *island growth* (Volmer-Weber growth), the material unit that is combined with metal and ligand is preferentially deposited on the ALD-grown material. Also, the growth mode may change randomly during film growth from two-dimensional growth to island growth and vise versa. This random growth mode is Stranski-Krastanov growth.

Growth behavior of ALD Sn phase was shown in figure 2. and the growth mode control was possible using in-situ gas treatment. As shown in figure2 (a), the growth mode of ALD Sn is *island growth*. In this case, Sn-Sn adatom cohesive force is stronger than surface adhesive force between Sn and Si_3N_4 substrate. The ALD Sn growth behavior was controlled by in-situ gas treatment and the growth mode was changed from island growth to random growth mode. This phenomenon may be explained by following two possible reasons. Before Sn precursor feeding to chamber, effective site for adsorption on substrate can be activated by in-situ gas treatment. In addition, steric hindrance can be minimized as a result of ligand exchange between Sn precursor and in-situ gas. As a result, the coverage of Sn monolayer increases in early stage of film growth and Sn adatom cohesive force is relatively reduced.

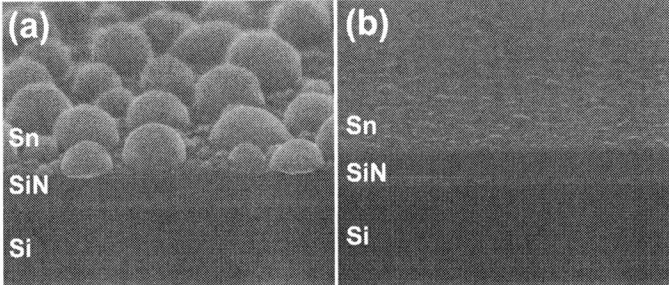

Figure 2. Growth behavior of ALD Sn single by SEM (a) without in-situ treatment, (b) with in-situ treatment

Stoichiometry of ALD SSS ternary alloy was controlled by the exposure ratio of Sb and Se precursor at a given SbSe and Sn sub-cycle ratio. (Sub-cycle ratio was fixed.) Sb/Se precursor exposure ratio was quantified by the feeding time of each precursors and the exposure ratio was varied from 0.5 to 8.0 as shown in figure 3. As a result, Sb and Se composition in ALD SSS film was controlled from 42.4% to 92.7%, 6% to 57% respectively. In addition, Sn composition slightly changed with Sb/Se exposure ratio from 0.6 to 3% because Sn incubation time to adsorb on Sb (or Se) terminated surface is different with Sb/Se exposure ratio at a fixed Sn precursor feeding time and sub-cycle.

Sb contents in ALD SSS films are increased and Se contents decreased when Sb/Se exposure ratio increase from 0.50 to 0.86. ALD grown materials competitively are adsorbed on substrate or precursor terminated surface. Generally, metal-ligand adsorption (including both physisorption and chemisorption) rate is proportional to the partial pressure and exposure time of precursor. When Sb/Se exposure ratio reaches 0.86, ALD SSS composition is saturated and slightly increased from 0.86 to 8.0 of Sb/Se exposure ratio. This tendency is one of the evidence of self-terminate reaction of ALD process. Riikka L. Puurunen explained the factors causing saturation behavior in ALD[17]. There are two factors causing the saturation of the surface with adsorbed species in a self-terminating gas-solid reaction. One is steric hindrance of the ligands and the other one is the number of reactive surface sites. The adsorption rate of Sb-ligands species linearly increase from 0.5 to 0.86 of Sb/Se exposure ratio and when reaches at 0.86, steric hindrance occurs and thus Sb-ligand species are no longer adsorbed even reactive surface sites exist.

Sn contents in ALD SSS films are increased from 0.6% to 3.0% at a range of 0.5-0.86 Sb/Se exposure ratio and decreased from 3.0% to 1.0% after 1.0 of Sb/Se exposure ratio. At a fixed Sn exposure and sub-cycle, Sn composition increases with decrease of Se composition from 0.5 to 0.75 Sb/Se exposure ratios. Self-terminate reaction of Sn precursor was shown from 0.75 to 1.0 Sb/Se exposure ratio and uniform Sn composition of 3% was shown. This phenomenon means that uniform doping of Sn is possible from 0.75 to 1.0 Sb/Se exposure ratio.

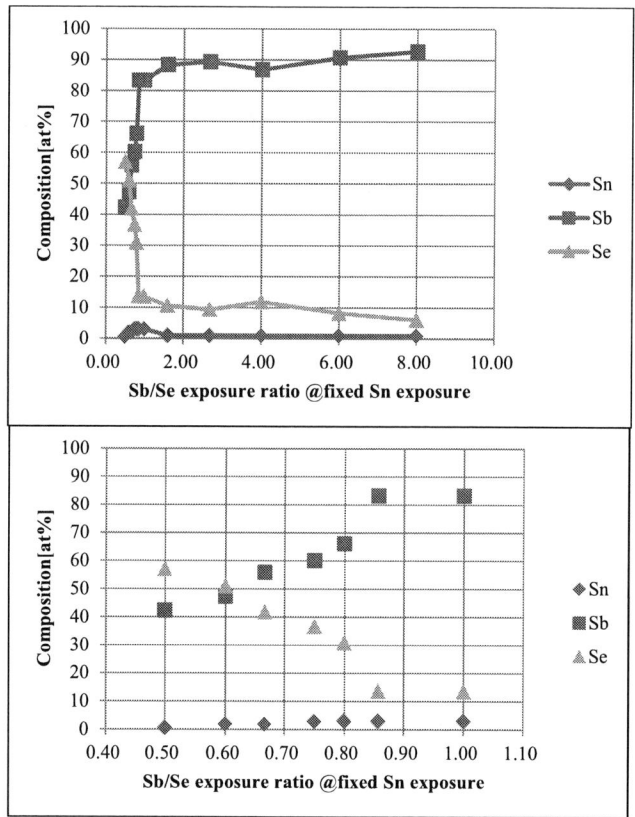

Figure 3. ALD SSS composition with Sb/Se exposure ratio at at fixed Sn exposure (feeding time and sub-cycle)

Characterization of ALD Sn-doped Sb-rich SbSe thin films was confirmed with 0.8 of Sb/Se exposure ratio. Cross-sectional and surface morphology images of as-deposited ALD SSS films are shown in figure 4. Flat and smooth surface morphology was obtained and RMS value which corresponds with a film roughness is 0.76nm. Film thickness was 260Å and confirmed by TEM shown in figure 5. As-deposited ALD SSS film shows nano-mixed phase and amorphous phase was confirmed by XRD. (Inset)

Figure 4. Surface morphology of as-deposited ALD SSS films (SEM)

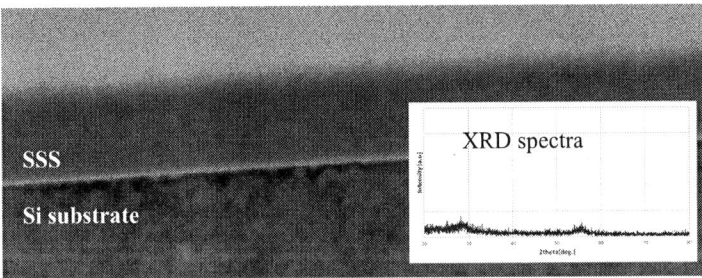

Figure 5 Cross-sectional image(TEM) and XRD spectra of as-deposited ALD SSS films

Laser induced phase transition behavior of as-deposited SSS film was investigated with laser static test. Contour map of reflectivity variation with laser power and laser pulse duration was shown in figure 6. Two distinct regions were discriminated by the reflectivity change. Crystallized region(I) and ablated region(II) were observed. However, crystallization is limited to particular section above laser power of 40mW and below 150ns pulse duration. The crystallization speed of ALD SSS film is 40ns at 70mW laser power.

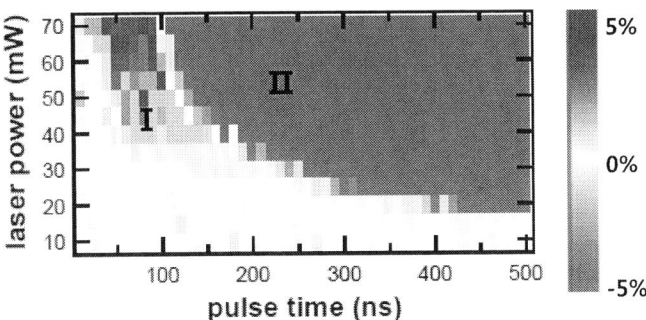

Figure 6. Contour map of reflectivity variation of as-deposited ALD SSS film with laser power and laser pulse duration.

In order to fabricate high-density PCRAM device with low power consumption, scaling down of cell size is required and phase change material should be filled in narrow hole. The gap-filling performance of ALD SSS films was evaluated with confined cell structure.(Figure 7.) Aspect ratio of the confined cell is 5.0 and the cell dimension is 15nm. Confined cell was completely filled with ALD SSS film and it shows excellent gap-filling performance without physical void or seam.

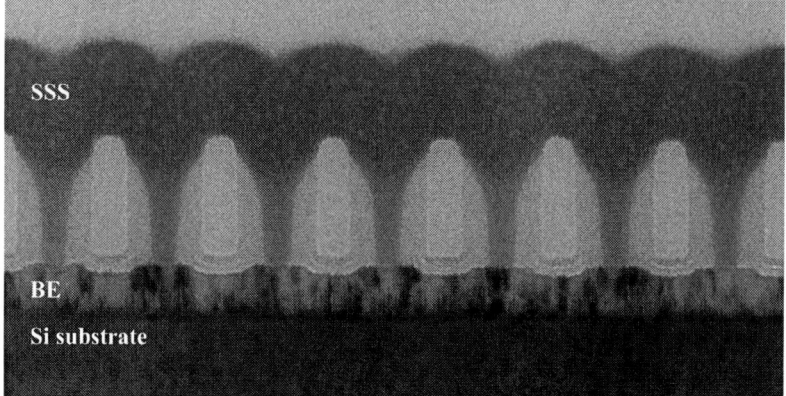

Figure 7. Gap-filling of ALD SSS film using confined cell structure. (Aspect ratio 5.0, Cell dimension of 15nm)

Conclusion

Sn-doped Sb-rich SbSe films were deposited by thermal ALD. Microstructure and growth behavior of ALD SSS thin films were investigated. Stoichiometry of ALD SSS ternary alloy was successfully controlled by the exposure ratio of Sb and Se precursor at a given SbSe and Sn sub-cycle ratio. As a result, Sb and Se composition in ALD SSS film was controlled from 42.4% to 92.7%, 6% to 57% respectively. In addition, Sn composition slightly changed with Sb/Se exposure ratio from 0.6 to 3%. As-deposited ALD SSS film shows amorphous phase and has very fast crystallization speed of 40ns. In addition, ALD SSS film shows excellent gap-filling performance for high performance PCRAM device.

References

1. C.M. Lee, D.S. Chao, Y.C. Chen, M.J. Chen, P.H. Yen, C.W. Chen, H.H. Hsu, W.H. Wang, W.S. Chen, F. Chen, T.C. Hsiao, M.J. Kao, M.J. Tsai, *International Symposium on VLSI Technology, Systems, and Applications* (2007)
2. T.J. Park, S.Y. Choi, M.J. Kang, *Thin Solid Films,* **515,** 5049 (2007)
3. E.M. Sanchez, E.F. Prokhorov, J.G. Hernandez, A.M. Galvan, *Thin Solid Films,* **471,** 243-247 (2005)
4. Y.K. Kim, M.H. Jang, K. Jeong, M.H. Cho, K.H. Do, D.H. Ko, H.C. Sohn, M.G. Kim, *Appl. Phys. Lett.,* **92,** 061910 (2008)
5. H. Seo, T.H. Jeong, J.W. Park, C. Yeon, S.J. Kim, S.Y. Kim, *Jap. J. Appl. Phys.* **Vol.39,** pp745-751 (2000)
6. S.M. Kim, J.H. Jun, D.J. Choi, S.K. Hong, Y.J. Park, *Jap. J. Appl. Phys.* **Vol.44,** No.6, ppL208-210 (2005)
7. B. Liu, Z. Song, J. Xia, S. Feng, B. Chen, *Thin Solid Fims,* **478** pp49-55 (2008)
8. N. Matsuzaki, K. Kurotsuchi, Y. Matsui, O. Tonomura, N. Yamamoto, Y. Fujisaki, N. Kitai, R. Takemura, K. Osada, S. Hanzawa, H. Moriya, T. Iwasaki, T. Kawahara, N. Takaura, M. Matsuoka, M. Moniwa, *Tech. Dig. - Int. Electron Devices Meeting,* **738,** (2005)
9. J. Feng, Z.F. Zhang, Y.Zhang, B.C. Cai, *J. Appl. Phys.,* **101,** 074502 (2007)
10. S.W. Ryu, J.H. Oh, J.H. Lee, B.J. Choi, W. Kim, S.K. Hong, C.S. Hwang, H.J. Kim, *Appl. Phys. Lett.,* **92,** 142110 (2008)
11. H.H. Shin, Y.S. Kang, A. Benayad, K.H. Kim, Y.M. Lee, M.C. Jung, T.Y. Lee, D.S. Suh, K.H.P. Kim, C.K. Kim, Y.H. Khang, *Appl. Phys. Lett.,* **93,** 021905 (2008)
12. R. Kojima, N. Yamada, *Jap. J. Appl. Phys..* **Vol40** pp. 5930-5937 (2001)
13. R.E. Simpson, D.W. Hewak, P. Fons, J. Tominaga, S. Guerin, B.E. Hayden, *Appl. Phys. Lett.,* **92,** 141921 (2008)
14. S.M. Yoon, N.Y. Lee, S.O. Ryu, K.J. Choi, Y.S. Park, S.Y. Lee, B.G. Yu, M.J. Kang, S.Y. Choi, M. Wuttig, *IEEE ELECTRON DEVICE LETTERS,* **VOL27,** NO. 6, pp 445-447 (2006)
15. K. Nakayama, K. Kojima, Y. Imai, T. Kasai, S. Fukushima, A. Kitagawa, M. Kumeda, Y. Kakimoto, M. Suzuki, *Jap. J. Appl. Phys..* **Vol42** pp. 404-408 (2003)
16. I.S. Kim, S.L. Cho, D.H. Im, E.H. Cho, D.H. Kim, G.H. Oh, D.H. Ahn, S.O. Park, S.W. Nam, J.T. Moon, C.H. Chung, *2010 Symposium on VLSI Technology Digest of Technical Papers* (2010)
17. Riikka L. Puurunen, *J. of Appl. Phys.,* **97,** 121301 (2005)

Rechargeable Li-ion Battery Anode of Indium Oxide with Visible to Infra-red Transparency

M. Osiak[1], W. Khunsin[2], E. Armstrong[1], T. Kennedy[3,4], C. M. Sotomayor Torres[2,5,6], K. M. Ryan[3,4], and C. O'Dwyer[1,4,7]

[1] Department of Chemistry, University College Cork, Cork, Ireland
[2] Catalan Institute of Nanotechnology, Campus UAB, Edifici CM3, Bellaterra, 08193 (Barcelona) Spain
[3] Department of Chemical and Environmental Sciences, University of Limerick, Limerick, Ireland
[4] Materials and Surface Science Institute, University of Limerick, Limerick, Ireland
[5] Catalan Institute for Research and Advanced Studies (ICREA), 08010 Barcelona, Spain
[6] Department of Physics, Universidad Autonoma de Barcelona, Campus UAB, 08193 Bellaterra, Spain
[7] Micro & Nanoelectronics Centre, Tyndall National Institute, Dyke Parade, Cork, Ireland

Unique bimodal distributions of single crystal epitaxially grown In_2O_3 nanodots on silicon are shown to have excellent IR transparency greater than 87% at 4 μm without sacrificing transparency in the visible region. These broadband antireflective nanodot dispersions are grown using a two-step metal deposition and oxidation by molecular beam epitaxy, and backscattered diffraction confirms a dominant (111) surface orientation. We detail the growth of a bimodal size distribution that facilitates good surface coverage (80%) while allowing a significant reduction in In_2O_3 refractive index. The (111) surface orientation of the nanodots, when fully ripened, allows minimum lattice mismatch strain between the In_2O_3 and the Si surface. This helps to circumvent potential interfacial weakening caused by volume contraction due to electrochemical reduction to indium, or expansion during lithiation. Cycling under potentiodynamic conditions shows that the transparent anode of nanodots reversibly alloys lithium with good Coulombic efficiency, buffered by co-insertion into the silicon substrate. These properties could potentially lead to further development of similarly controlled dispersions of a range of other active materials to give transparent battery electrodes or materials capable of non-destructive in-situ spectroscopic characterization during charging and discharging.

Introduction

The marked increase in portable electronic device sales together with huge demand for flat screen high-definition televisions (HDTVs) are the main driving forces behind the need for batteries and continued research into various materials and forms for transparent conducting oxides (TCOs) and similar coatings (1). Among TCOs, materials such as indium oxide (IO), tin oxide (TO) or tin-doped indium oxide (ITO) (2–6) and emerging alternatives such as graphene and Cu or Ag NWs for example (6,7), have been a consistent focus of research interest where transparency in a useful visible range is matched by sheet resistances below 10 $\Omega/\square$ (9) ITO is the TCO used most often and its applications vary from thin film transistors (10–12) to transparent contact in solar cells (5). Low sheet resistances are typically required for thin-film solar cells and the solar photon flux-weighted optical transparency of ITO on glass is about 80%. The battery however, a key component in the majority of portable electronics, has only very recently been demonstrated as a transparent device (13), and there is room for the development of true see-through charge storage materials (14–17). Their metallic properties cause most TCO's to be reflective in the infrared and for most TCO's a trade-off exists between transparency, conductivity, and sheet resistance for thin films (18). In line with this, transparency in battery electrodes gives the opportunity for *in-situ* and non-destructive diagnostic analysis of material changes during battery operation. This research could also allow the possibility of investigating kinetics of intercalation mechanisms and the influence of certain lithiated phases of TCO materials on transparency and conductivity.

Here, In_2O_3 {111}-oriented crystalline nanodot dispersions have been successfully grown from an MBE deposition of an In layer and subsequent oxidation at elevated temperature. The method results in unique areal and size dispersions of nanodots varying in size from hemispherical 2 nm dots to larger, faceted ~500 nm crystals, on the Si current collectors. Angle-resolved transmittance measurements confirm that the deposits maximize transparency in the infra-red, while maintaining characteristic transparency in the visible with a beneficial reduction in resistivity and sheet resistance; this overcomes the transparency limitations for In_2O_3 nanomaterials by index matching with air through a unique size dispersion. The nanodots form as $In@In_2O_3$ core-shell crystals, and form a stable solid electrolyte interface (SEI) layer and reversibly alloy with lithium allowing them to function as visible-to-IR transparent, visibly antireflective Li-ion battery electrodes. The approach shown here is straightforward and scalable and may be applied to the fabrication of high quality optoelectronic, electronic and sensor devices. Moreover, it could introduce visible-to-IR transparent conducting TCOs that reversibly store (electro) chemical charge, and also develop non-destructively, optically addressable materials and interfaces for *in-situ* monitoring of electrochemical processes.

Experimental

Before growth on silicon and glass substrates, the respective surfaces were cleaned using a standard RCA process. After rinsing, a second treatment in a H_2O_2:HCl:H_2O (1:1:5) solution was used to remove metallic and organic contamination. For evaporation of the In sources, a home-built MBE high-vacuum chamber with a distinct effusion cell for In together with an electron-beam evaporator was designed in cooperation with MBE-Komponenten GmbH As detailed Fig. 1, a uniform layer of In metal was deposited at a rate of 0.1 Å s^{-1} at a substrate temperature of 400°C, with precise control over the nominal thickness.

Surface morphologies and the chemical composition of the nanostructured dispersions were investigated by electron microscopy using a Hitachi SU-70 SEM with an Oxford-50mm^2 X-Max detector for energy dispersive X-ray analysis and Oxford Instruments Nordlys EBSD detector with HKL Channel 5 acquisition software. The size distribution of the nanodots was analysed using ImageJ (19).

X-ray photoelectron spectroscopy was acquired using a Kratos Axis 165 monochromatized X-ray photoelectron spectrometer equipped with a dual anode (Mg/Al) source. Survey spectra were captured at as pass energy of 100 eV, step size of 1 eV, and dwell time of 50 ms. The core level spectra were an average of 10 scans captured at a PE of 25 eV, step size of 0.05 eV, and dwell time of 100 ms. The spectra were corrected for charge shift to the C 1s line at a binding energy of 284.9 eV. A Shirley background correction was employed, and the peaks were fitted to Voigt profiles.

Variable angle spectroscopic ellipsometry (VASE) was performed using a J. A. Woollam Co., Inc. M-2000U variable angle spectroscopic ellipsometer over a wavelength range of 300 to 900 nm. Reflectance measurements were carried out in a Bruker FT-IR spectrometer IFS66/V on nanodot samples and ITO on glass. Different configurations of beam splitters, detectors and sources were used to cover the infrared (5 μm) to visible ranges. For angular resolved measurements, a NIR512 Ocean Optics spectrometer was used as a detector in a home-built reflectance/transmittance setup using a collimated Xenon arc lamp as a light source.

To investigate the electrochemical insertion (alloying) and removal of Li, cyclic voltammetry measurements were carried out in a 3-electrode setup using a Multi Autolab 101 potentiostat, using Li as both counter and reference electrodes. All potentials, unless otherwise stated, are relative to Li^+/Li. Custom build swagelock-type cells were used with counter and active material electrode separated by a polypropylene separator soaked in 1 mol dm^{-3} solution of $LiPF_6$ in EC:DMC at a 50:50 v/v ratio. The electrode was cycled at a scan rate of 0.5 mV/s. Afterwards, the electrode was carefully washed in acetonitrile and a 10^{-4} mol dm^{-3} solution of acetic acid to remove the electrolyte residue.

Results and Discussion

Epitaxial growth of In_2O_3 nanodots

MBE deposition of indium and subsequent oxidation in ambient air allows the formation of a specific size dispersion of oxide crystals after In growth, as shown in Fig. 1a. The dispersion consists of larger crystals interspersed with a high density of very small (~2-5 nm) nanodots (Fig. 1b). Some of the larger crystals have clearly developed facets generally growing in a deviated hexagonal shape (Fig. 1b). High resolution SEM images of the nanodots show that a number of small crystallites are found on the top surfaces of the large crystals (Fig. 1b).

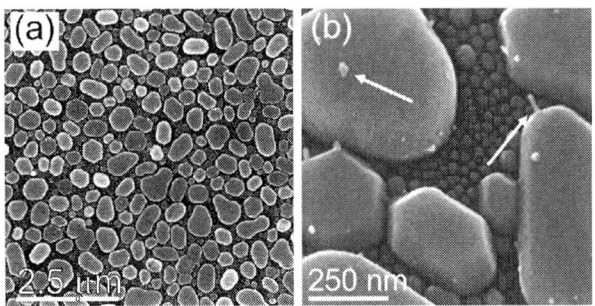

Figure 1. (a) SEM image showing the final epitaxial In_2O_3 nanodot dispersion on Si. (b) Tilted SEM image showing small, hemispherical nanodots interspersed between larger crystals. Arrows indicate small crystallites growing hierarchically on the top surfaces and nanowires growing from the edges of crystals.

The initial formation of a dewetted liquid In 'layer' comprising a high density of metallic nanodots (maintained in a liquid state on a substrate heated to 400 °C), and the progressive nature of their deposition allows hierarchical nanodot seeds to form on the high energy facet edges of the larger crystals. In some cases we observe subsequent growth of long, straight In_2O_3 nanowires (see Fig. 1b) with lengths reaching hundreds of nanometres and diameters not exceeding the diameter of hierarchical dots.

Electron backscattered diffraction from the terminating surfaces of both faceted and non-faceted nanodots was used to quantify their epitaxial relationship to their substrate and also their relative orientational distribution. The measurements were taken at 70° tilt (Fig. 2a) and a pole plot of the nanodot texture orientation distribution (Fig. 2b) was formed by monitoring the Kikuchi diffraction patterns from the top surface lattice planes of the nanodots shown in Figs 2c and d; the growth orientations from 3D crystal symmetry are visible as diffraction 'paths' in orientation-space. The measurements confirm a dominant $\{111\}$ surface termination for the nanodots. Interestingly, for both faceted and non-faceted crystals, their terminating planes are near-identical, as are their overall heights of ~50 nm, see Fig. 2a.

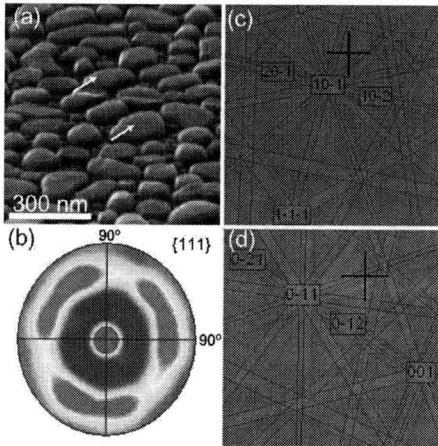

Figure 2. (a) SEM image of the In_2O_3 nanodots. Arrows indicate the location of the points at which the EBSD pattern was recorded. (b) Pole figure showing the relative orientation distribution of $\{111\}$ termination of the nanodots. (c-d) Kikuchi band overlays recorded from the regions indicated in (a). The cross on each image indicates the orientation of the planes at their measurement point.

Being extremely sensitive to tilt or variations in the top surface of the crystals, the corresponding pole plot shown in Fig. 2b were acquired to map the distribution of textures around major growth directions. The texture pole plot is centered around the $\{111\}$ directions. It is clear from the texture distributions in the in Fig. 2b, that the particles grow with horizontal hexagonal $\{111\}$ planes, parallel to the (100) substrate of the silicon wafer. The In_2O_3 nanodots were epitaxially deposited as metal nanodot seeds and subsequently oxidized in air and EBSD analysis confirms that their oxidation to In_2O_3 nanodots results in a final single crystal structure with the $\{111\}$ growth planes parallel to the substrate. The growth rate perpendicular to $\{111\}$ places is comparatively slower than $\{110\}$ and $\{100\}$ planes. As a result, growth in lateral direction progresses faster than in the vertical, which we find regardless of the degree of crystal faceting. Additionally, lattice mismatch ($f = (a_f - 2a_{sub})/2a_{sub}$ where a_f and a_{sub} are the lattice constants in the growth plane of $In_2O_3(111)$ and $Si(100)$, respectively) of In_2O_3 on silicon is only 1.13% resulting in low strain at the nanodot-substrate interface (20). This allows for minimizing of any additional strain placed on the deposit due to the electrochemical reduction from In_2O_3 to metallic In and subsequent volumetric expansion accompanying electrochemical Li insertion.

The composition of the MBE nanodots was determined using XPS and EDX. Figure 3a shows the In3d and O1s core-level photoelectron emission spectra of the nanodots. Core-level emission corresponding to In $3d_{5/2}$ and In $3d_{3/2}$ were observed at 444.34 eV and 452.03 eV (referenced to the C1s core-level of 284.9 eV) indicative of In_2O_3. The peak at 444.34 eV shows hyperfine levels, one at 443.9 eV from In(0) and at 445.1 eV related to In $3d_{5/2}$ from In_2O_3. Core-level emission from O 1s was composed of two spectral bands at 531.2 eV and 529.6 eV, which can be deconvoluted into three components consistent with In_2O_3. The signal at 531.2 eV is attributed to lattice oxygen, while that at 529.6 eV stems from some $In(OH)_3$, which is known to form from exposure of In_2O_3 to water vapour. Corresponding EDX maps of In and O (shown in Figs 3b-e) corroborate oxide composition of the nanodots.

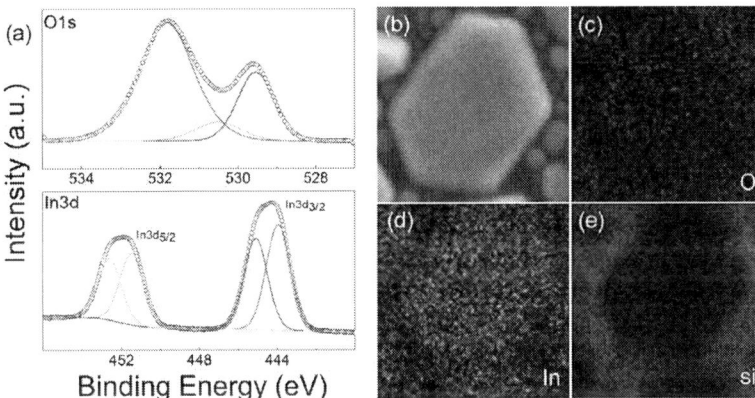

Figure 3. X-ray core-level photoelectron spectra of (a) O 1s and In 3d of the In_2O_3 nanodot dispersion. (b-e) EDX maps of In_2O_3 nanodot showing distributions of oxygen, indium and silicon respectively.

Enhanced IR transparency of In_2O_3 nanodot dispersion.

Angle resolved transmission measurements of the nanodot dispersions and an ITO thin film were determined and are summarised in Fig. 4. The position of the plasma frequency is indicated by ω_p undergoes a red shifts and the reflectance of the In_2O_3 nandots dispersion and the ITO thin film decreases with angle near their respective plasma frequencies, shown in Fig. 4a. The reflectance decreases substantially after the plasma frequency. Nanoparticle layers offer excellent visible-infrared transmission, and also antireflection properties, as seen in the optical images in Fig. 4b and c. The red-shifting of the entire angle-resolved spectrum for the nanodots dispersions is also seen in Fig. 4a, where at visible wavelengths the transmission varies from 55% at near incidence and at 40°, and importantly, remains 87% transparent at wavelengths up to 4 μm (Fig. 4a) (20).

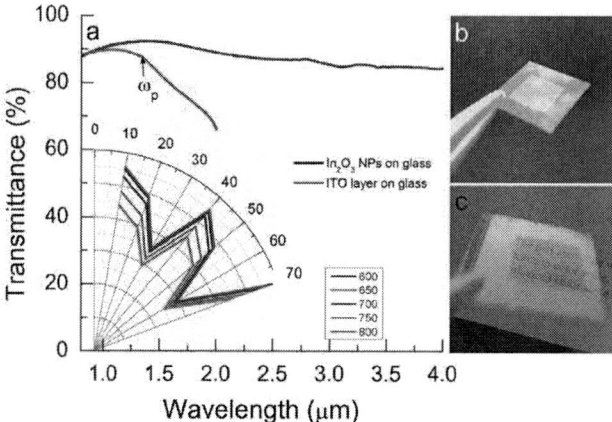

Figure 4. (a) Transmittance of the In_2O_3 nanodot dispersions and that of an ITO thin film of similar nominal thickness. (*Inset*) Polar plot of the angle-resolved transmission of the nanodots at visible wavelengths. (b,c) Optical images of the In_2O_3 nanodots showing antireflection characteristics in the visible range.

Reversible electrochemical Li-insertion

The ability of the In_2O_3 nanodot dispersions to reversibly intercalate or alloy Li, and its insertion and removal potentials, were examined using cyclic voltammetry. Figure 5a shows the cyclic voltammetric response of the $Li|Li^+$-electrolyte$|In_2O_3(111)|Si(100)$ system. For this cell the cathodic process included the insertion of Li into In_2O_3 to form a Li-In alloy (charging) and the anodic process follows Li extraction or dealloying (discharging). A related process is known to occur for Li alloying with Sn (21), but there are limited investigations of Li insertion in to In-containing materials (22). During the first negative scan, two weak irreversible peaks appear at 1.2 V and 0.8 V from the reduction of In_2O_3 to In^0. Once reduced from In_2O_3 to In^0, the indium is never oxidized again in the potential range examined. Zhou *et al*. (23) have shown reoxidation after cycling the anode to upper potentials greater than 3.5 V, a voltage window typical of cathode materials. We cycled the anode in 0 - 2.5 V, a potential window below oxidation potential of In^0 (2.7 V vs. Li^+/Li) (24).

A large reversible peak appears at 0.4 V from the alloying process of Li insertion into In^0. The extent of this reaction, indicated by measured current is found to reduce with increasing cycle number. The reversible Li insertion-removal process occurs in a voltage window of $0.4 - 0.7$ V are the reversible processes described by $zLi^+ + ze^- + In \leftrightarrow Li_zIn$ $(0 < z \leq 4.33)$. For $Li|Li^+$-electrolyte$|In_2O_3(111)|Si(100)$, buffering of polarization effects is provided by the Si current collector, which can accommodate the highest Li storage capacity of all anode materials (25,26).

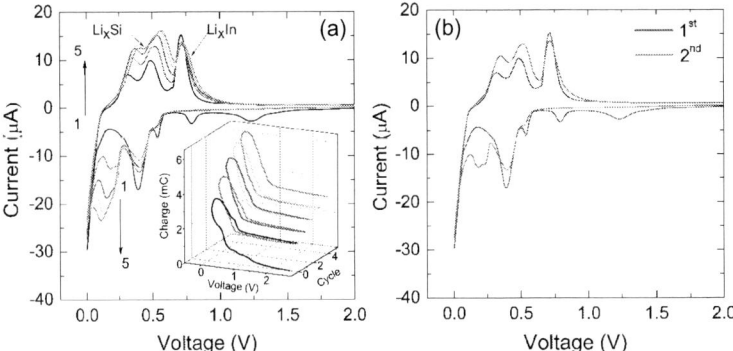

Figure 5. (a) Cyclic voltammetry of In_2O_3 nanodot electrodes between $0.0 - 2.5$ V. Inset shows the corresponding integrated charge vs. voltage curve for 5 cycles. (b) Cyclic voltammograms for first and second cycle highlighting the SEI layer formation.

The free surface area (~20%) between neighbouring particles allows Li to be co-inserted into the silicon current collector, which is indicated by the existence of two additional peaks in the anodic part of the insertion reaction, at 0.32 V and 0.5 V. Those peaks relate to the removal of lithium from silicon (25). Figure 5b shows the first 2 cycles of this system, and we note that the reduction of In_2O_3 and related alloying processes dominate over Li_xSi phase formation and insertion of lithium into silicon in the first cycle. This co-insertion into the active material and current collector equilibrates after the second cycle. The increasing rate of Li-Si formation can be attributed to an activation effect (26) linked to lithiation-induced volumetric expansion that causes cracking and the exposure of unreacted material to the electrolyte. Successive cycling then allows more lithium to intercalate into silicon, providing a degree of stress buffering for the In alloying process without requiring carbon, conductive additives or polymeric binders. The rate of alloying and dealloying, insertion and removal are consistently balanced in each cycle (Fig. 5a, inset), and apart from charge associated with SEI formation and reduction to In^0, negligible charge fading is found for *all* processes in all subsequent cycles.

During co-insertion into the active material and current collector, both of which are reversible, the variation in volumetric changes and accompanying effects is considered. The SEM images in Figs 6a and b show the condition of the electrode surface before and after cycling. The brightness of the secondary electron emission stems from a reduced conductivity of the Li_xIn phase. As the nanodots are epitaxial, their adhesion to the substrate is excellent, and lithium insertion is not likely to occur directly under each nanodot, unless they are extremely small. In this case, we note that some of the smallest nanodots are removed from the substrate, but this occurs when their diameter is less than the change in volume of the near surface of the silicon. The molar volume of In^0 is a factor of 2.45 less than the In_2O_3 and by comparison to the size reduction observed, it is clear that no significant volume change effects occur in stable In^0 nanodots faceting related to the structure of In_2O_3 is also lost during electrochemical reduction to the pure metal.

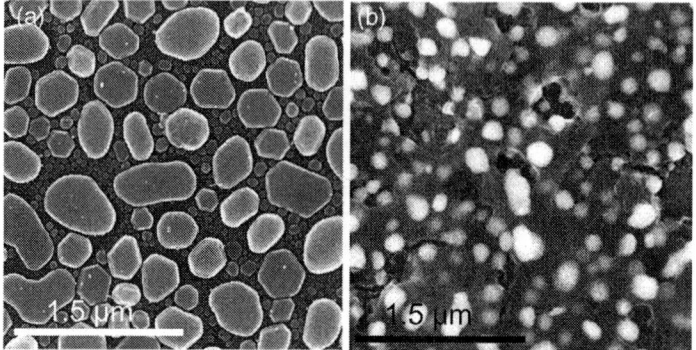

Figure 6. SEM images showing nanodots (a) and (b) after electrochemical cycling.

Conclusions

The unique size dispersion of In_2O_3 nanodots prepared by MBE deposition of indium and subsequent oxidation in air at elevated temperature, has allowed the development of a Li-ion battery electrode with enhanced IR transparency without sacrificing electrical conductivity, and lithium co-insertion processes with high Coulombic efficiency that results in stable cycling and charge storage. The In_2O_3 nanodots show bimodal size distribution confirming a two-step epitaxial growth mechanism, and good surface coverage with unique shape and (111) crystalline orientation. The nanodot dispersions were successfully shown to reversibly alloy with lithium after reduction to metallic indium; the specific size distributions allow reversible lithium co-insertion with a silicon current collector as well as the active material on the surface. Moreover, the specific size offers excellent antireflective properties and enhanced transparency reaching ~87% at 4 μm, potentially allowing for further development of transparent battery electrodes or the possibility for *in-situ* non-destructive spectroscopic monitoring of structural and electrochemical processes.

Acknowledgements

MO and EA acknowledge the support of the Irish Research Council under awards RS/2010/2170 and RS/2010/2920. WK and CMST acknowledge support from the Spanish MINECO projects ACPHIN (FIS2009-10150) and TAPHOR (MAT2012-31392), the CONSOLIDER project nanoTHERM (CSD2010-00044) and the Catalan AGAUR grant 2009-SGR-150. COD acknowledges support from Science Foundation Ireland under award no. 07/SK/B1232a and from UCC Strategic Research Fund. The authors thank F. Laffir and C. Dickinson for assistance in XPS and EBSD measurements.

References

1. D. S. Ginley and C. Bright, *MRS Bull.*, **25**, 15–18 (2011).
2. C. H. Chiu, P. Yu, C. H. Chang, C. S. Yang, M. H. Hsu, H. C. Kuo, and M. a Tsai, *Opt. Express.*, **17**, 21250–6 (2009).
3. P.-C. Chen et al., *ACS nano*, **3**, 3383–90 (2009).

4. N. R. Armstrong, P. A. Veneman, E. Ratcliff, D. Placencia, and M. Brumbach, *Acc.Chem. Res.*, **42**, 1748–57 (2009).

5. C. O'Dwyer, M. Szachowicz, G. Visimberga, V. Lavayen, S. B. Newcomb, and C. M. S. Torres, *Nat. Nanotechnol.*, **4**, 239–44 (2009).

6. H. K. Yu, W. J. Dong, G. H. Jung, and J.-L. Lee, *ACS nano*, **5**, 8026–32 (2011).

7. P.-C. Hsu, H. Wu, T. J. Carney, M. T. McDowell, Y. Yang, E. C. Garnett, M. Li, L. Hu, and Y. Cui, *ACS nano*, **6**, 5150–6 (2012).

8. K. S. Kim et al., *Nature*, **457**, 706–10 (2009).

9. J.-Y. Lee, S. T. Connor, Y. Cui, and P. Peumans, *Nano Lett.*, **8**, 689–92 (2008).

10. D. Lin, H. Wu, R. Zhang, and W. Pan, *Nanotechnology*, **18**, 465301 (2007).

11. S. Kim, S. Ju, J. H. Back, Y. Xuan, P. D. Ye, M. Shim, D. B. Janes, and S. Mohammadi, *Adv. Mater.*, **21**, 564–568 (2009).

12. K. Samedov, Y. Aksu, and M. Driess, *Chem. Mater.*, **24**, 2078–2090 (2012).

13. Y. Yang, S. Jeong, L. Hu, H. Wu, S. W. Lee, and Y. Cui, *P. Natl. Acad. Sci.*, **108**, 13013–8 (2011).

14. Z. Wu et al., *Science*, **305**, 1273–6 (2004).

15. T. M. Barnes, X. Wu, J. Zhou, a. Duda, J. van de Lagemaat, T. J. Coutts, C. L. Weeks, D. a. Britz, and P. Glatkowski, *Appl. Phys. Lett.*, **90**, 243503 (2007).

16. M. Zhang, S. Fang, A. a Zakhidov, S. B. Lee, A. E. Aliev, C. D. Williams, K. R. Atkinson, and R. H. Baughman, *Science*, **309**, 1215–9 (2005).

17. H.-K. Kim, D.-G. Kim, K.-S. Lee, M.-S. Huh, S. H. Jeong, K. I. Kim, and T.-Y. Seong, *Appl. Phys. Lett.*, **86**, 183503 (2005).

18. P. D. C. King and T. D. Veal, *J. Phys.: Condens. Matter*, **23**, 334214 (2011).

19. C. a Schneider, W. S. Rasband, and K. W. Eliceiri, *Nat. Methods*, **9**, 671–675 (2012).

20. D. M. Follstaedt, *Appl. Phys. Lett.*, **62**, 1116 (1993).

21. J. S. Chen and X. W. (David) Lou, *Mater. Today*, **15**, 246–254 (2012).

22. H. Geaney, T. Kennedy, C. Dickinson, E. Mullane, A. Singh, F. Laffir, and K. M. Ryan, *Chem. Mater.*, **24**, 2204–2210 (2012).

23. Y. Zhou, H. Zhang, M. Xue, C. Wu, X. Wu, and Z. Fu, *J. Power Sources*, **162**, 1373–1378 (2006).

24. J. Vanhees, J. P. Francois, and L. C. Van Poucke, *J. Phys. Chem.*, **85**, 1713–1718 (1981).

25. C. K. Chan, H. Peng, G. Liu, K. McIlwrath, X. F. Zhang, R. a Huggins, and Y. Cui, *Nat. Nanotechnol.*, **3**, 31–5 (2008).

26. C. K. Chan, R. N. Patel, M. J. O'Connell, B. a Korgel, and Y. Cui, *ACS nano*, **4**, 1443–50 (2010).

27. B.-S. Lee, S.-B. Son, K.-M. Park, J.-H. Seo, S.-H. Lee, I.-S. Choi, K.-H. Oh, and W.-R. Yu, *J. Power Sources*, **206**, 267–273 (2012).

CHAPTER 3

POROUS SEMICONDUCTORS AND SEMICONDUCTING OXIDES

ECS Transactions, 53 (6) 65-79 (2013)
10.1149/05306.0065ecst ©The Electrochemical Society

Cessation of Porous Layer Growth in n-InP Anodised in KOH

Robert P. Lynch,[a,b] Nathan Quill,[a,b] Colm O'Dwyer,[c,d] Monika Dornhege,[e]
Harm H. Rotermund [e,f] and D. Noel Buckley [a,b]

[a] Department of Physics and Energy, University of Limerick, Limerick, Ireland
[b] Materials and Surface Science Institute, University of Limerick, Limerick, Ireland
[c] Applied Nanoscience Group, Dept. of Chemistry, University College Cork, Ireland
[d] Micro & Nanoelectronics Centre, Tyndall National Inst., Lee Maltings, Cork, Ireland
[e] Surface Imaging Group, Fritz-Haber-Institute of the Max-Planck-Society, Department of Physical Chemistry, Berlin, Germany
[f] Surface Reaction Imaging Group, Department of Physics and Atmospheric Science, Dalhousie University, Halifax, Nova Scotia, Canada

Anodisation of n-InP in KOH results in the formation of porous layers with a finite thickness. We propose the reason for the cessation of porous etching is the formation of insoluble precipitates within the pores. Electron micrographs of mature porous layers show significant precipitates within the porous structure. An *in-situ* microscopy study of the surface of InP electrode during anodisation reveals the formation of a layer on the surface. This layer emerges from a point on the surface and quickly spreads across it. A likely source of this layer is the spreading of precipitation from the etch-products saturated solution within the porous layer. However, as we explain, once a complete porous layer has formed, there should be no significant increase in mass transport requirements through the porous network, leaving the exact mechanism of the precipitation unclear.

Introduction

Although the formation of porous silicon in HF has long been known,[1-4] the discovery of visible luminescence from porous silicon [5] sparked the surge in research interest in the formation of porosity in other semiconductors. The list of semiconductors that can now be rendered porous electrochemically includes germanium,[6-8] GaP,[9-11] InP,[12-18] GaAs,[19-23] GaN,[24-26] and many others. A range of different pore morphologies can be obtained in these semiconductors by variation of electrolyte type and concentration,[14,27] carrier concentration and substrate orientation,[28,29] as well as the current density or potential used to form the porous structure.[30] A number of theories have been proposed [2-4,31-34] to explain the plethora of pore morphologies that have been observed with various semiconductor/electrolyte combinations but, so far, no one theory has been able to explain all observations.

In our group, we have demonstrated the formation of porous InP in KOH electrolytes in the concentration range 1-17 mol dm^{-3}.[18,35] We have previously shown that pores emerge from pits in the electrode surface[36] and grow and branch along the <111>A crystallographic directions, forming tetrahedral porous domains.[37] However, unlike

65

porous InP formed in acidic solutions[12-14], in KOH the pore propagation spontaneously halts both in linear-potential-sweep (LPS) and potentiostatic experiments resulting in a rapid decrease in current density. In this paper we will describe the cessation of this pore propagation and examine the possible causes of this cessation.

Experimental

Wafers were monocrystalline, sulfur-doped, n-type indium phosphide (n-InP) grown by the liquid-encapsulated Czochralski (LEC) method and supplied by Sumitomo Electric. They were polished on one side and had a surface orientation of (100) and a carrier concentration of ~5 × 10^{18} cm^{-3}. To fabricate working electrodes, wafers were cleaved into coupons along the natural {011} cleavage planes. Ohmic contact was made by alloying indium to the back of each coupon; the back and the cleaved edges were then isolated from the electrolyte by means of a suitable varnish. The electrode area was typically 0.2 cm^2. Prior to immersion in the electrolyte, the working electrode was immersed in a piranha etchant (3:1:1 $H_2SO_4:H_2O_2:H_2O$) for 4 minutes and then rinsed with deionized water.

Anodisation was carried out in 5 mol dm^{-3} aqueous KOH by linear potential sweep (LPS) experiments at a scan rate of 2 − 2.5 mV s^{-1}. A conventional three-electrode cell configuration was used, employing a platinum counter electrode and a saturated calomel electrode (SCE) to which all potentials are referenced. A potentiostat interfaced to a Personal Computer (PC) was employed for cell parameter control and for data acquisition. All experiments were carried out at room temperature in the absence of light unless otherwise stated.

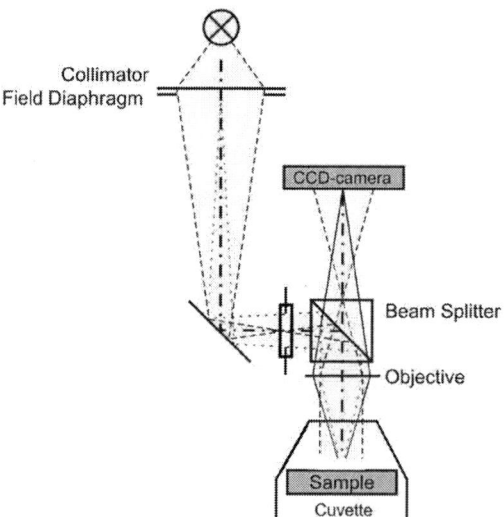

Figure 1. Schematic of optical setup for imaging the electrode surface *in situ* using Köhler illumination at an incident angle of 0°.

In certain experiments low intensity Köhler illumination was used to facilitate *in-situ* microscopy of the electrode surface. We have described the layout of the experiment and the technique in detail elsewhere[38-40]. During potential scans the surface was imaged via optical microscopy with a CCD camera. Great care was taken to synchronize the video sequences and the current versus time curves. A sketch of the optical setup is shown in Fig. 1.

Cleaved {011} cross-sections of the coupons were examined *ex situ* using a Hitachi S-4800 field emission scanning electron microscope (FE SEM) operating at 5 kV. Crystallographic directions were determined by reference to the primary $(0\overline{1}\overline{1})$ and secondary $(0\overline{1}1)$ flats of the supplied wafers.

Results and Discussion

LPS of n-InP in 5 mol dm^{-3} KOH

Anodisation of n-InP samples by linear potential sweep (LPS) was carried out in 5 mol dm^{-3} KOH at a scan rate of 2.5 mV s^{-1}. The resulting linear-sweep voltammogram (LSV) is shown in Fig. 2c. Marked on the LSV are the pitting potential E_{pit}, the position of the first current peak (P_1) and the position of the second current peak (P_2). After E_{pit} the current rapidly increases until it reaches the current of P_1. During this period, isolated porous domains can be observed on SEM cross section (*e.g.* Fig. 2a). After a short decay in current due to the merging of separate porous domains into a complete porous layer (*e.g.* as shown in Fig. 2b),[41] the current begins to increase with potential again, at an approximately linear rate. This approximately linear increase in current is due to the thickening of a fully formed porous layer. Eventually, the current peaks again at P_2 and decays rapidly thereafter. This rapid decay in current is characteristic of passivation of the electrode against further porous layer growth. From this point on the porous layer ceases etching. A further increase of the potential will eventually lead to a change in etching mechanism, initially resulting in the undercutting of the porous layer and etching of the perimeter of the sample[38] followed by electropolishing of the remaining electrode surface.

Similar InP porous layers can be formed in acidic electrolytes such as HCl. Such layers typically exhibit crystallographically oriented porous layers which can readily extend over 30 µm into the InP surface[42] and current-line oriented porous layers which can extend over 100 µm into the InP surface[43]. The possible reasons for the cessation of porous layer growth in InP in KOH will be explored in the ensuing sections.

Formation of Precipitates within the Porous Layer

Porous layers which have been grown to the point of cessation are typically observed to have significant deposits within their porous structure. To study the appearance of such deposits within the pores, a series of samples were anodised by LPS to various peak potentials and the resulting porous layers were examined by scanning electron microscopy (SEM). The top left image in Fig. 3 shows an LSV of an InP electrode in 5 mol dm^{-3} KOH. The four points marked A through D on the LSV indicate the upper potentials of LPSs performed on samples under similar conditions. Figures 3a-3d show

SEM micrographs of samples swept to points A, B, C and D, on the LSV on the top left of the figure. Image (a) shows an SEM micrograph of the sample that had its potential swept to point A (0.34 V). A typical incomplete porous layer is seen with empty 'clean' pores seen throughout. Images (b1) and (b2) show the upper and lower sections, respectively, of a porous layer grown to point B (0.385 V) in the LSV. The upper section is near the electrode surface, and the lower section is near the bulk InP. The pores in the upper section are again typical empty pores. However, in the lower section of the porous layer a small amount of material can be seen as grainy, roughened deposits in the pores in image (b2). Images (c1) and (c2) show the upper and lower sections, respectively, of a porous layer grown to point C (0.42 V). At this point, even the upper section is showing some deposits within the pores, with large lumps seen to be clogging up some of the pores. In the lower section, deposits can now be seen in almost every pore. Finally, images (d1) and (d2) show the upper and lower sections respectively of a porous layer grown to point D (0.5 V). The upper section of the layer is again filled with many lumps which have a diameter roughly equal to the width of the pore. The lower section in this instance is so completely filled with material that it is difficult to distinguish between deposit-filled pores and bulk InP.

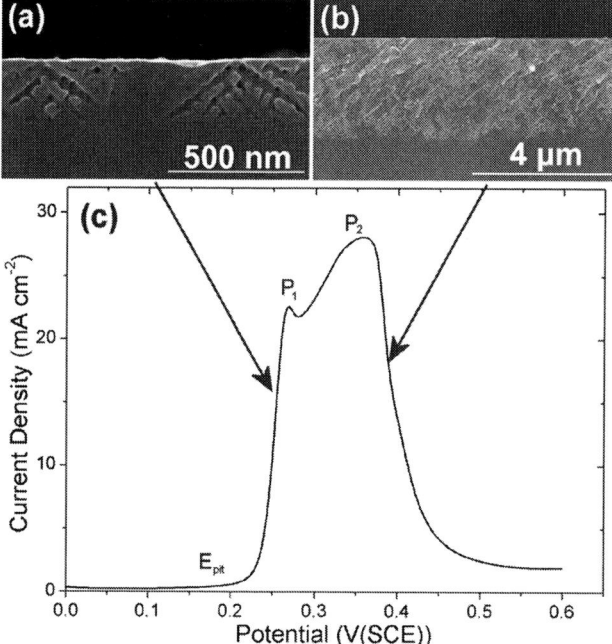

Figure 2. (a) SEM image of two separate porous domains about to merge together, (b) SEM image of a complete porous layer (Note: the surface oxide layer was removed during preparation for SEM) and (c) a LSV of an n-InP sample subjected to an LPS from 0 to 0.6 V at 2.5 mV s^{-1} in 5 mol dm^{-3} KOH. The points marked on the LSV are E_{pit}, the potential at which etch pits appear on the electrode surface, and P_1 and P_2, the first and second current peaks, respectively.

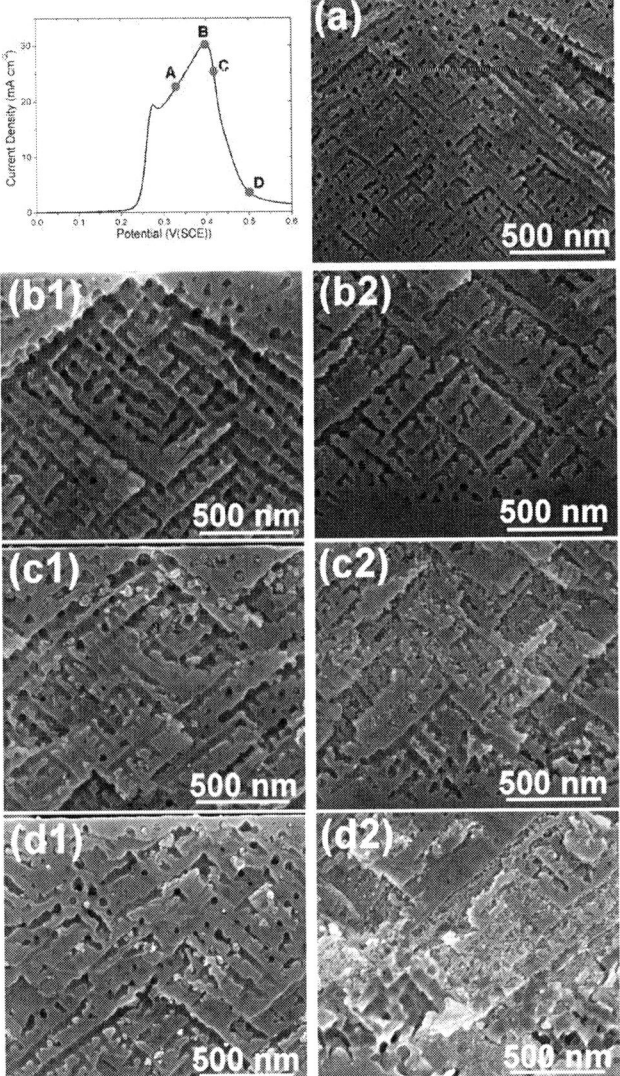

Figure 3. SEM micrographs of InP electrodes which have been subjected to LPSs to potentials indicated on the LSV on the top left of this figure. Images (a), (b), (c) and (d) show porous layer grown to point A (0.33 V), B (0.40 V), C (0.42 V) and D (0.50 V) in the LSV, respectively, where image 1 and 2 of (b), (c) and (d) show the upper and lower section of the respective layers. Image (d) corresponds to a fully grown porous layer.

The micrographs in Fig. 3 show a clear trend. After P_2, the pores begin to fill up with deposits. This indicates a likely correlation between the end of porous layer etching and the appearance of these deposits. In Image (d2) the in filling is so complete as to block the path of the electrolyte through the pores. It seems likely that the formation of these precipitates results in the observed decrease in current and the termination of porous layer etching. The precipitates are always more abundant near the porous-bulk interface which suggests that its source may be at or near the pore tips. This indicates that the deposits are likely to be composed of at least some of the reaction products of the InP dissolution.

In-situ Observation of Precipitates on the Electrode Surface

Figure 4 shows the process of etching InP, as viewed with a CCD camera, under low intensity Köhler illumination. Several stages can be distinguished during the LPS as discussed previously[38]. During the first 90 s the current increases slowly with increasing applied potential, with no resulting change to the electrode surface (*e.g.* Fig. 4a). Just before 0.2 V, the current begins to increase rapidly and continues to increase until it reaches 44.6 mA cm^{-2}. As the current reaches this first peak the light intensity from the surface reaches a maximum (*e.g.* Fig. 4b) due to the formation of a thin near surface layer of dense InP that is separated from the bulk InP by a continuous porous layer (similar to that shown at A in Fig. 5). Coinciding with the second current peak the electrode surface darkens (*e.g.* Fig. 4c). The current then decreases and simultaneously a bright region appears and grows finger-like into the dark area (as is evident from Figs. 4d and 4e). Once this bright region has spread over the surface the light intensity is relatively unaffected by the anodisation process despite the current continuing to increase until the end of the experiment.

The increase in light intensity observed in Fig. 4b has been explained previously by us to be due to constructive interference[38]. Where reflection occurs at a transition from high to low refractive index, the reflected light experiences a phase shift of half a wavelength. Therefore, a very thin layer of InP ($\eta_{InP} \approx 3.7$) will produce two reflected rays that will destructively interfere if it is between electrolyte ($\eta_e \approx 1.4$) and a porous layer that contains electrolyte similar to the bulk electrolyte (*i.e.* $\eta_{pl} < 3.7$). We refer to this situation where the second layer has a higher refractive index than that of the first and third layers as a Type I system. In such a system, if the layer thickness is a quarter of the wavelength of the light constructive interference will occur. Such a condition is met for the thickness of the dense near surface layer.

In a similar manner a very thin layer of InP will produce two reflected rays that will constructively interfere if it is between electrolyte ($\eta_e < \eta_{InP}$) and a porous layer that contains electrolyte of refractive index greater than the that of InP ($\eta_{pII} > 3.7$). We refer to this situation where the refractive index of the third layer is greater than that of the second and the refractive index of the second is greater than that of the first as a Type II system. In such a system, if the layer thickness is a quarter of the wavelength of the light, destructive interference will occur. The required layer thickness, Δt, for constructive and destructive interference maxima can therefore be calculated for both Type I and Type II systems as follows:

TABLE 1. Conditions for constructive and destructive interference.

Layer Type:	Type I: $\eta_e < \eta_d > \eta_{pI}$	Type II: $\eta_e < \eta_d < \eta_{pII}$
Destructive Interference:	$\Delta t = \dfrac{n\lambda_d}{2}$	$\Delta t = \dfrac{\lambda_d}{4} + \dfrac{n\lambda_d}{2}$
Constructive Interference:	$\Delta t = \dfrac{\lambda_d}{4} + \dfrac{n\lambda_d}{2}$	$\Delta t = \dfrac{n\lambda_d}{2}$

where $n = 0,1,2,3\ldots$, λ_d is the wavelength of the light in the medium of the second layer. For In_2O_3 the wavelength of the light is $\lambda_d = \lambda_o/\eta_{ox} \approx 252$ nm ($\lambda_o = 530$ nm, In_2O_3 refractive index, $\eta_{ox} \approx 2.1$[44]), and for InP it is $\lambda_d = \lambda_o/\eta_{InP} = 142.6$ nm (InP refractive index, $\eta_d = 3.72$[45]). (The refractive index of the of 5 mol dm^{-3} KOH is $\eta_e = 1.38$.[46]).

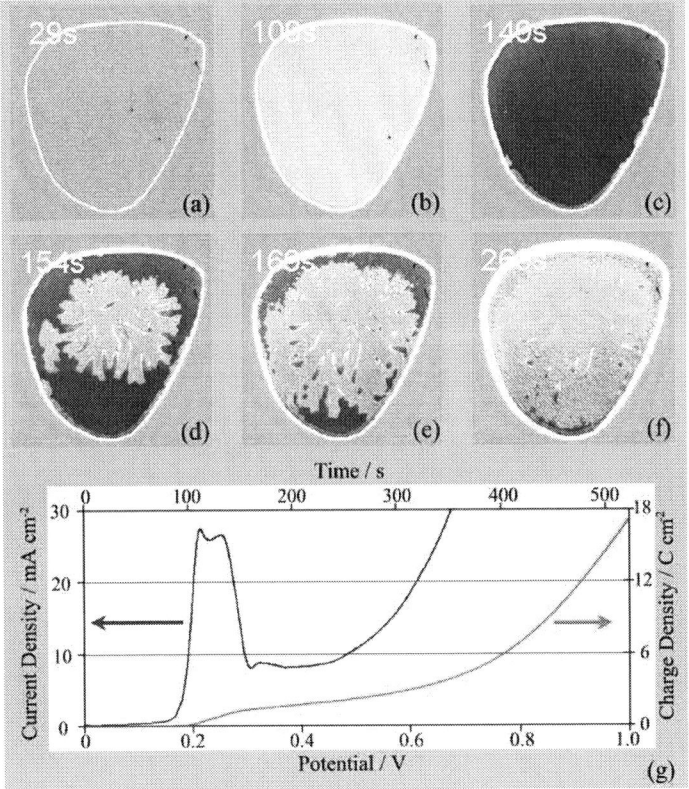

Figure 4. (a-f) *In-situ* microscopy images of the electrode surface during anodisation of InP by LPS. The number of seconds corresponds to the points in the voltammogram (LSV) and charge versus potential plot shown in (g). The experiment was carried out under a low intensity of light from 0 to 1 V (SCE) in 5 mol dm^{-3} KOH at 2 mV s^{-1}.

At approximately the same time as the second current-peak (see Fig. 4c), the reflected light from the electrode surface darkens. This outcome could be the result of destructive interference caused by the near-surface layer thinning or thickening or the order of the refractive indexes changing from a Type I to a Type II system. The near-surface layer decreases to 20 nm in some SEM images, accounting for a decrease in light intensity but not for the decrease that is observed. (The light intensity decreases to less than that of the light reflected from the surface at the beginning of the experiment.) However, the type of interference could switch from constructive to destructive interference if the order of the refractive indices switched from a Type I to a Type II system, *i.e.* if the refractive index of the material within the pores became greater than that of InP.

As the current decreases after the second current peak, interference fringes appear on and spread out over the electrode surface. These fringes could be caused by the growth of a porous oxide-layer on the surface with constructive or destructive interference occurring depending on the thickness of the porous layer and the order of the refractive indices, *i.e.* whether it is a Type I or Type II system, as shown in Table 1. The propagation of these fringes is impeded by scratches on the surface supporting the conclusion that the fringes are related to phenomena on the electrode surface and not to some sub-surface region of the porous structure. This can be seen in Fig. 5 where the black lines (scratches) divide the fringe regions apart.

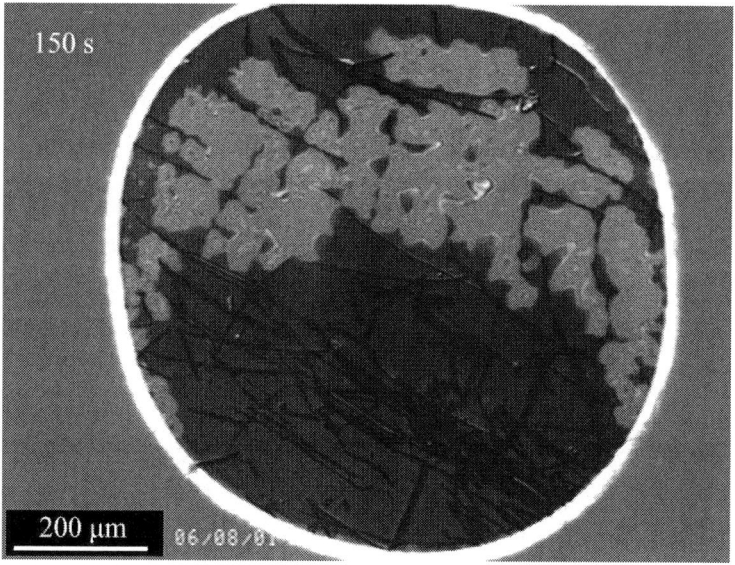

Figure 5: *In-situ* optical microscopy image of the electrode surface at 150 s during an LPS from 0 to 1 V (SCE) in 5 mol dm^{-3} KOH at 2 mV s^{-1}. The surface was scratched prior to etching and these scratches are impeding the progress of the fringe pattern across the electrode surface.

After this decrease in current and the formation of the finger-like pattern of interference fringes, porous layer growth halts and the appearance of the surface changes from being smooth to rough (*e.g.* Fig. 4e to Fig. 4f). This effect starts quickly but then slows down to leave only some small regions that contain interference fringes, but these too eventually disappear. One possibility is that the pits on the electrode surface become blocked, restricting the diffusion of electrolyte leading to the precipitation of etch products and eventually the stopping porous layer growth. Oxide layers are observed on the surface of fully grown porous layers. In Fig. 6 an oxide layer, approximately 0.72 μm in thickness, can be seen on the surface of the electrode (at C). Energy dispersive X-ray spectroscopy (EDX), X-ray photoelectron spectroscopy (XPS) and electron diffraction studies have shown this oxide to be composed of In_2O_3[47,48]. In_2O_3 is known to be insoluble in all but the most acidic of solutions[49] so it is not surprising that it precipitates during the formation of porous InP in KOH. Assuming a refractive index close to that of In_2O_3, a rough calculation shows that its thickness is sufficient to account for destructive and constructive interference patterns formed on the surface. Furthermore the thickness of the porous layer (at B) is unchanged from the thickness of the porous layer measured by SEM (not shown) for samples anodised by LPS to lower potentials, *i.e.*, the thickness of the porous layer does not increase after the potential of the second peak in current. However, even though the growth of the porous layer stops after this point in the LSV, the current increases. The increase in current has been shown previously to be due to the formation of a trench around the perimeter of the exposed region of the electrolyte[38] (which is observed as a bright outline in *in-situ* microscopy images, *e.g.* Fig. 4f).

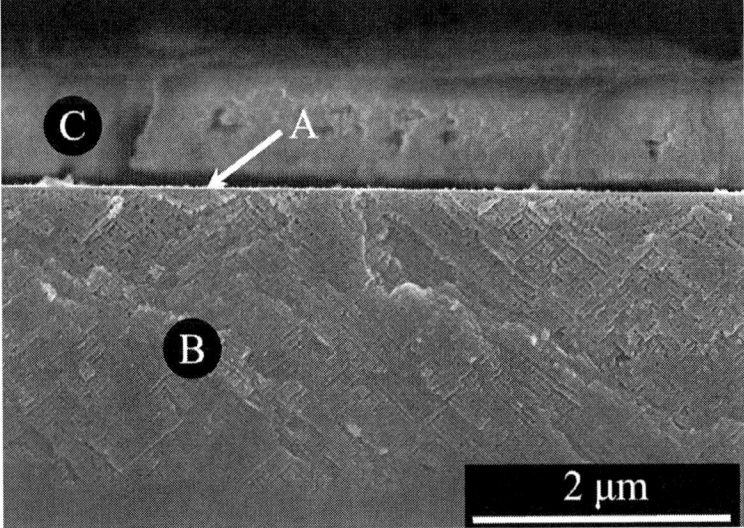

Figure 6: SEM image of porous layer cross-sections (at B) of InP etched in 5 mol dm^{-3} KOH at 2 mV s^{-1} from 0 V to 1 V (SCE) (500 s). The near-surface layer and a thick oxide layer are visible (at A and C, respectively).

Discussion of Diffusion of Ions through the Porous Layer

The observation of precipitates within the pores (Fig. 3) appearing after P_2 coupled with the observation of a layer spreading on the surface (Fig. 4) after P_2 would seem to indicate that it is the build up of precipitates in the pores as etching progresses that leads to the cessation of porous layer growth. However, the mechanism by which such precipitates initially form and build up is not obvious.

As a porous domain emerges from a single pit in the electrode surface, the amount of material that needs to be transported through its surface pit increases rapidly as the domain expands. This is due to each surface pit branching into two pores along the two <111>A directions which point down into the substrate, followed by the subsequent branching of the two new pore tips. All of these new pore tips rely on the lone surface pit as the only pathway to transport etch products out of the porous structure and to transport fresh electrolyte into the porous structure. The surface pit then, is the main mass transport bottleneck during the formation of porous InP. The amount of mass that must be transported through the pit would be expected to increase rapidly as an individual porous domain expands and the number of actively etching pore tips relative to each surface pit increases. However, once a certain porous layer thickness has been reached (determined by both the final pit density and the rate of pit formation), the individual porous domains merge into a continuous layer. Individual domains can no longer expand laterally, and will only expand deeper into the substrate. Since the pore density and pore width have not been observed to change with porous layer thickness for crystallographically oriented pore etching in KOH, the number of pore tips per porous domain should stay relatively constant after a complete layer has formed. This is because the domain's active etching area can no longer increase by expanding laterally, and the pore tips associated with that domain should always occupy the same area. Therefore once a complete layer has formed, the rate at which material is being transported through each surface pit has reached a constant value *i.e.* the ratio of the number of pore tips to the number of surface pits is now constant. (Note: in LPS experiments, the current continues to increase in an approximately linear fashion after domain merging has completed, placing a further strain on mass transport processes. However, potentiostatic experiments also exhibit similar behavior during the cessation of porous layer growth.)

The expansion and merging of domains is illustrated in Fig. 7 which shows a two dimensional view of the change in active etching area for the porous etching process as individual domains converge to form a continuous porous layer. The red line representing the active etching area is considerably shorter for two merged domains than it is for two isolated domains. The difference in concentration of a particular ion from one side of a surface pit to the other must increase rapidly until the domains have merged. Similarly the change in concentration of this ion between the bulk electrolyte and the pore tips initially increases rapidly as the domains expand. After domain merging, this difference in concentration continues to increase at a much lower rate, due to the thickening of the layer, and in the case of LPS experiments, the approximately linear increase in current density. However, compared to the rapid increase in concentration difference that is observed before domain merging occurred, the increase after domain merging is much less significant.

If mass transport limitations are the cause of the cessation of porous layer etching, it is much more likely for pore propagation to cease before domain merging has occurred than after. This is what happens in lower carrier concentration electrodes[41]. Such electrodes show a single peak in their LSV indicating that the cessation of porous layer etching had begun to occur before a complete porous layer had formed. Higher carrier concentration samples exhibit two peaks in their LSVs and these were shown to be related to domain merging and the cessation of porous layer growth, respectively[41]. In many of these LSVs, the amount of charge passed through the electrode after domain merging is many times the amount of charge passed through the electrode before domain merging. This indicates that domain merging occurs at a shallow layer thickness and that layer thickness increases by a significant amount before the cessation of porous layer etching begins. If mass transport difficulties within the porous network result in the build-up and precipitation of etch products within the pores, then this precipitation should occur before the domains have merged and an etch front of constant area has been achieved. Once the domains have merged and a complete layer has formed, one would not expect the concentration of the dissolved etch products within the pores to increase by a significant amount, and the porous layer should be able to thicken continuously without much difficulty.

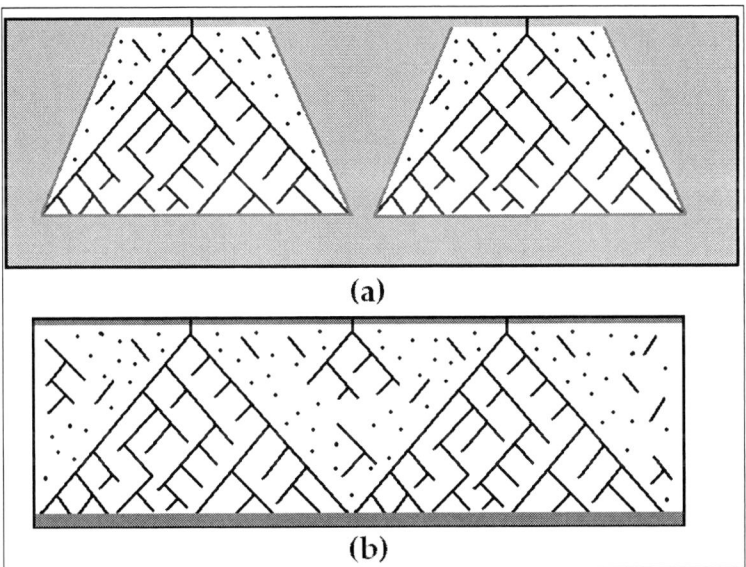

Figure 7. Schematic diagram of the variation in active etching area as (a) individual porous domains expand and (b) as a complete porous layer thickens. The grey areas represent bulk InP. The porous regions have a white background with pores being shown as black lines and dots. The red line indicates the active etching area as seen from this cross section. Clearly the active etching area decreases once a complete layer has formed.

Nevertheless, porous layer etching does cease after a few micrometers, and significant amounts of precipitates are observed in the porous structure. One possible mechanism for the termination of porous layer etching would involve the precipitation being initiated in a small number of pores. Some domains will be bigger than others or a region of a domain may expand in a region where it is not competing with other domains (e.g. grow a little faster than neighboring domains). In such situations the concentration of products in a pore may approach saturation much more rapidly. Where this leads to precipitates in one of these domains, they may block a number of diffusion pathways that are connected to other pores in that domain causing further precipitation. The precipitation of solids would result in the expansion of the material within the pores. The saturated solution should therefore be pushed throughout the domain and out through the surface pit forming more precipitates as it spreads. These precipitates might restrict diffusion of material through the surface pits of neighboring domains leading to the saturation of etch products near the pore tips of these domains. Furthermore once etching has ceased in a domain neighboring domains will expand laterally, rapidly increasing the concentration of products near their pore tips. It follows that once etching ceases in one domain the cessation of etching spreads to neighboring domains.

Such a mechanism can explain what is seen in *in-situ* optical microscopy experiments as shown earlier in Fig. 4: No significant changes occur on the electrode surface between the formation of a complete layer (P_1) and the beginning of the cessation of porous layer growth (P_2) in LPS experiments. However, the electrode surface darkens as $P2$ is being approached, possibly due to the solution in the pores becoming more and more saturated with etch products changing the refractive index of the solution. Also, just after P_2, interference patterns start to appear at a small number of points on the surface and spread out over the whole surface (Figs. 4d and 4e), corresponding to a solution saturated with etch products being pushed out from the pores due to the formation of precipitates within the porous layer. Since this layer of solution has a different refractive index to both the electrode and the bulk electrolyte interference fringes are observed, corresponding to the different thicknesses of the layer. Furthermore as precipitates adhere to the electrode surface and the remaining etch products diffuse into the bulk electrolyte the reflection from the surface will change from specular to diffuse as observed in Fig. 4f. This is most likely the formation of the oxide layer that is seen in SEM images of electrodes with fully grown porous layers (*e.g.* at C in Fig. 6). It also follows, that regions where specular reflection is maintained (*e.g.* in Fig. 4f) correspond to domains where pores continue to propagate. However, the domains of such pores will be virtually unrestricted in how far they can expand latterly. Thus, such domains will become filled with precipitates, stopping further pore propagation and leading to reflection from the surface of the electrode becoming diffuse.

The fact that a layer of different refractive index emerges from just a small number of points on the electrode surface just after P_2, indicates that it may be precipitation in localized regions of the porous layer that initiates a domino effect of almost simultaneous cessation of the InP porous layer thickening, infilling of pores with precipitates and formation of an oxide layer on the electrode surface. That is, the expansion of the material within the pores leads to the transfer of etch-product saturated solution to the electrode surface. This solution restricts the diffusion of reactants to the pore tips and products from the porous layer. The concentration of the trapped products within the porous structure would increase quickly, leading to precipitation near almost all of the pore tips explaining why precipitates are seen in most of the pores near the bottom of a fully etched layer.

The above mechanism relies on the random precipitation of some etch products for its initiation. This may seem at odds with the relatively consistent layer depths that are observed from experiment to experiment. However, the large number of pores involved (typically $>2\times10^{10}$ cm^{-2} for crystallographically oriented pore layers in KOH) should result in consistent behavior, even if events are defined by probabilities. For example, surface pits are said to form at defect sites. These defect sites are randomly distributed across the electrode surface, but measurements of pit density are consistent from sample to sample, and the shift in pitting potential from sample to sample is similar to the shift seen in the potential at P_2 from sample to sample. This demonstrates that when such large numbers are involved, even probabilistic events occur consistently.

Conclusions

Anodisation of InP in KOH results in the formation of porous layers of finite thickness. This thickness limit is likely due to the low solubility of etching products in alkaline solutions. SEM images of such porous layers reveal that they contain significant precipitates and that their electrode surface is often capped by an oxide layer < 1 μm thick, which is most likely composed of In_2O_3.

In-situ optical observation of the electrode surface during etching shows what appears to be the spreading of a layer over the whole electrode surface from points on the surface. The formation of this layer coincides with the cessation of porous layer thickening and samples that have undergone LPSs as far as the corresponding potentials exhibit significant infilling of pores and covering of the electrode surface by precipitates in SEM images.

These findings indicate that it may be precipitation of etch products in just a small number of pores that starts the cessation of porous etching via a domino effect. This domino effect could propagate due to the lateral expansion of domains that neighbor domains which have ceased etching and/or the restriction of diffusion through the surface pits due to the formation of a surface layer of precipitates above domains that have ceased etching. Both of these occurrences lead to the saturation of products near the pore tips of domains that neighbor domains that have ceased etching, leading to the ceasing of etching also at these pore tips. The spreading of these precipitates eventually blocks the surface pits of all domains leading to increased mass transport difficulties in the porous layer and the formation of precipitates initially near the tips of each pore (where the concentration of products is greatest) and eventually along almost all of the length of each pore. Indeed, even where a small number of domains continue to etch the lateral expansion of such domains will lead saturation of etch products that precipitate, stopping all pore propagation.

Acknowledgements

Two of the authors, R.P. Lynch and N. Quill, would like to thank the Irish Research Council (IRC) for postgraduate scholarships to perform this research.

References

1. A. Uhlir, *Bell Syst. Tech. J.*, **35**, 333 (1956).
2. M. I. J. Beale, J. D. Benjamin, M. J. Uren, N. G. Chew and A. G. Cullis, *J. Cryst. Growth*, **73**, 622 (1985).
3. R. L. Smith and S. D. Collins, *J. Appl. Phys.*, **71**, R1 (1992).
4. X. G. Zhang, *J. Electrochem. Soc.*, **151**, C69 (2004).
5. L. T. Canham, *Appl. Phys. Lett.*, **57**, 1046 (1990).
6. S. Miyazawa, K. Sakamoto, K. Shiba and M. Hirose, *Thin Solid Films*, **255**, 99 (1995).
7. M. Sendova-Vassileva, N. Tzenov, D. Dimova-Malinovska, M. Rosenbauer, M. Stutzmann and K. V. Josepovits, *Thin Solid Films*, **255**, 282 (1995).
8. S. Bayliss, Q. Zhang and P. Harris, *Appl. Surf. Sci.*, **102**, 390 (1996).
9. B. H. Erne, D. Vanmaekelbergh and J. J. Kelly, *J. Electrochem. Soc.*, **143**, 305 (1996).
10. P. Schmuki, D. J. Lockwood, H. J. LabbeÂ´ and J. W. Fraser, *Appl. Phys. Lett.*, **69**, 1620 (1996).
11. J. Wloka, K. Mueller and P. Schmuki, *Electrochem. Solid-State Lett.*, **8**, B72 (2005).
12. A. Hamamatsu, C. Kaneshiro, H. Fujikura and H. Hasegawa, *J. Electroanal. Chem.*, **473**, 223 (1999).
13. T. Takizawa, S. Arai and M. Nakahara, *Jpn. J. Appl. Phys.*, **33**, L643 (1994).
14. P. Schmuki, L. Santinacci, T. Djenizian and D. J. Lockwood, *Phys. Stat. Sol., A*, **182**, 51 (2000).
15. Z. Weng, A. Liu, Y. Sang, J. Zhang, Z. Hu, Y. Liu and W. Liu, *J. Porous Mater.*, **16**, 707 (2009).
16. Z. Weng, W. Zhang, C. Wu, H. Cai, C. Li, Z. Wang, Z. Song and A. Liu, *Appl. Surf. Sci.*, **256**, 2052 (2010).
17. A. M. Goncalves, L. Santinacci, A. Eb, I. Gerard, C. Mathieu and A. Etcheberry, *Electrochem. Solid-State Lett.*, **10**, D35 (2007).
18. C. O'Dwyer, D. N. Buckley, D. Sutton and S. B. Newcomb, *J. Electrochem. Soc.*, **153**, G1039 (2006).
19. G. Oskam, A. Natarajan, P. C. Searson and F. M. Ross, *Appl. Surf. Sci.*, **119**, 160 (1997).
20. M. M. Faktor, D. G. Fiddyment and M. R. Taylor, *J. Electrochem. Soc.*, **122**, 1566 (1975).
21. P. Schmuki, J. Fraser, C. M. Vitus, M. J. Graham and H. S. Isaacs, *J. Electrochem. Soc.*, **143**, 3316 (1996).
22. I. M. Tiginyanu, V. V. Ursaki, E. Monaico, E. Foca and H. Foll, *Electrochem. Solid-State Lett.*, **10**, D127 (2007).
23. A. M. Gonçalves, L. Santinacci, A. Eb, C. David, C. Mathieu, M. Herlem and A. Etcheberry, *Phys. Stat. Sol., A*, **204**, 1286 (2007).

24. D. J. Diaz, T. L. Williamson, I. Adesida and P. W. Bohn, *J. Vac. Sci. Technol., B*, **20**, 2375 (2002).
25. A. P. Vajpeyi, S. Tripathy, S. J. Chua and E. A. Fitzgerald, *Physica E*, **28**, 141 (2005).
26. F. K. Yam, Z. Hassan, L. S. Chuah and Y. P. Ali, *Appl. Surf. Sci.*, **253**, 7429 (2007).
27. N. Quill, R. P. Lynch, C. O'Dwyer and D. N. Buckley, *Abstracts of Joint 222nd Electrochemical Society Meeting and Fall Meeting of the Electrochemical Society of Japan, Symposium D7: Pits and Pores 5: A Symposium in Honor of David Lockwood,* Honolulu, Hawaii, USA, 7-12 October, abstract 2420 (2012).
28. N. Quill, C. O'Dwyer, R. Lynch and D. N. Buckley, *ECS Trans.*, **19**, 295 (2009).
29. S. Ronnebeck, J. Carstensen, S. Ottow and H. Foll, *Electrochem. Solid-State Lett.*, **2**, 126 (1999).
30. S. Langa, I. M. Tiginyanu, J. Carstensen, M. Christophersen and H. Foll, *Electrochem. Solid-State Lett.*, **3**, 514 (2000).
31. T. Unagami, *J. Electrochem. Soc.*, **127**, 476 (1980).
32. V. Lehmann and H. Foll, *J. Electrochem. Soc.*, **137**, 653 (1990).
33. V. Lehmann, *J. Electrochem. Soc.*, **140**, 2836 (1993).
34. J. Cartensen, M. Christophersen and H. Foll, *Mat. Sci. Eng., B*, **69-70**, 23 (2000).
35. R. P. Lynch, C. O'Dwyer, D. N. Buckley, D. Sutton and S. Newcomb, *ECS Trans.*, **2**, 131 (2006).
36. C. O'Dwyer, D. N. Buckley, D. Sutton, M. Serantoni and S. B. Newcomb, *J. Electrochem. Soc.*, **154**, H78 (2007).
37. R. P. Lynch, C. O'Dwyer, D. Sutton, S. B. Newcomb and D. N. Buckley, *ECS Trans.*, **6**, 355 (2007).
38 R. P. Lynch, M. Dornhege, P. S. Bodega, H. H. Rotermund and D. N. Buckley, *ECS Trans.*, **6**, 331 (2007).
39 M. Dornhege, C. Punckt, J. L. Hudson and H. H. Rotermund, *J. Electrochem. Soc.*, **154**, C24 (2007).
40. C. Punckt, M. Bölscher, H. H. Rotermund, A. S. Mikhailov, L. Organ, N. Budiansky, J. R. Scully and J. L. Hudson, *Science*, **305**, 1133 (2004).
41. N. Quill, C. O'Dwyer, R. Lynch and D. N. Buckley, *ECS Trans.*, **19**, 295 (2009).
42. E. Spiecker, M. Rudel, W. Jäger, M. Leisner and H. Föll, *Phys. Stat. Sol., A*, **202**, 2843 (2005).
43. L. Santinacci, A. M. Goncalves and A. Etcheberry, *ECS Trans.*, **6**, 323 (2007).
44. K.N. Rao, and S. Kashyap, *Surf. Rev. and Lett.* **13**, 221 (2006)
45. D.E. Aspnes and A.A. Studna, *Phys. Rev. B* **27**, 985 (1983)
46. CRC Handbook of Chemistry and Physics 86th ed.: *Physical Constants of Inorganic Compounds,* p. 4-65, D. R. Lide, Editor-in-Chief, CRC Press, New York (2005).
47. N. Quill, *Formation of Nanoporous InP in KOH and KCl: Electrochemistry and Electron Microscopy*, PhD Thesis, University of Limerick, Limerick (2012).
48. C. O'Dwyer, *Anodic Formation and Characterisation of Porous InP in KOH Electrolytes*, PhD Thesis, University of Limerick, Limerick (2003).
49. P. J. Durrant and B. Durrant, *Introduction to advanced inorganic chemistry*, p. 586, Wiley, New York (1970).

80

TiO₂ Nanotubes Formed in Aqueous Media: Relationship between Morphology, Electrochemical Properties and the Photoelectrochemical Performance for Water Oxidation

Próspero Acevedo-Peña and Ignacio González

Departamento de Química, Universidad Autónoma Metropolitana-Iztapalapa. Av. San Rafael Atlixco 186 C.P. 09340. Ciudad de México, México.

The effect of voltage and electrolyte composition on the formation of TiO_2 nanotube films over its morphology, semiconductor properties and photoelectrochemical performance for water oxidation was studied. TiO_2 nanotube films were formed by potentiostatic anodization in $0.1\,M\,H_2SO_4/0.05\,M\,HF$ and $0.1\,M\,H_2SO_4/0.05\,M\,NH_4F$ during 1 h at four different voltages. SEM images showed a more significant chemical attack when NH_4F was employed, evidenced by larger tube diameters and thinner films. The semiconductor properties were altered by the nature of the counter-ion utilized during film growth, obtaining films with a higher density of donor and less positive flat band potential in NH_4F electrolytes. The films formed in NH_4F showed higher photopotentials and photocurrent related to the modification of their morphology and semiconductor properties, attaining a maximum at formation voltage of 15 V.

Introduction

The anodization of Ti substrates to grow TiO_2 nanotubes in fluoride containing media, is a process that has been gaining increasingly importance as a result of its flexibility and simplicity of formation, high ordered structure, directional pathway for electron transport and the paramount performance showed by these films in different applications such: photocatalysis (1), photoelectrochemical solar cells (2), water splitting (3), among others (4, 5). Particularly, the water oxidation has drawn attention of different researchers since it is the half-reaction occurring over the anode (TiO_2 film) during photoelectrochemical water splitting (3, 6-9).

TiO_2 nanotubes can be obtained by anodization in aqueous electrolytes as well as in organic media containing fluoride ions (1-10). The latter one is the most commonly used to fabricate films for photoelectrochemical applications due to the simplicity to control the morphology (tube diameter and length) through operational variables. TiO_2 nanotubes formed in fluoride electrolytes are typically short due to the high chemical etching produced by fluoride ions during their growth (10). However, these films are highly ordered, adherent and require lower imposed voltage than those formed in organic media.

Different studies have tried to correlate the photoelectrochemical performance of TiO_2 nanotube films with its morphological characteristics (e.g. tube length, pore size and porosity) (8, 11), leaving aside the influence of the electronic properties (12-14), which are closely related to the methodology and the operational variables employed to grow the film (7, 12-17), and has shown to alter the photoelectrochemical performance of the TiO_2 films in different applications. Therefore, it is necessary to figure out how the electronic properties of TiO_2 nanotube films are being modified during anodization.

In our research group different *in-situ* studies have been carried to study the growth of nanoporous/nanotubular TiO_2 films by anodization in fluoride containing electrolytes, employing electrochemical impedance spectroscopy (18-20). This fundamental work have contributed to understand the relationship between processing variables and, properties and morphology of the obtained films. In acid aqueous electrolytes, varying the fluoride counter-ion from H^+ (HF) to NH_4^+ (NH_4F), causes a larger fluoride ion insertion and the formation of TiO_2 nanotube films with larger pore diameter and roughness (20). These modifications will definitely alter the electronic properties of the TiO_2 nanotube films; however, it is uncertain whether they will impact directly on the electronic properties of the films, after being subjected to a heat treatment, which is normally applied to the as-formed film for using in a photoelectrochemical cells.

Thus, the purpose of this study is to evaluate the modifications induced by the formation voltage and the fluoride counter-ion employed during anodization in $0.1\,M\,H_2SO_4/0.05\,M\,HF$ (labeled as HF) and $0.1\,M\,H_2SO_4/0.05\,M\,NH_4F$ (labeled as NH_4F), over the morphology, semiconductor properties and photoelectrochemical performance for water oxidation.

Experimental

The TiO_2 nanotube films were obtained by potentiostatic anodization in a two-electrodes cell during 1 h at different formation voltages (10 V, 15 V, 20 V and 25 V), employing two acid aqueous electrolytes: $0.1\,M\,H_2SO_4/0.05\,M\,HF$ (labeled as HF) and $0.1\,M\,H_2SO_4/0.05\,M\,NH_4F$ (labeled as NH_4F). Pt (99.99% Alfa Aesar) was employed as counter electrode placed at 2 cm from the Ti foil (99.95% Alfa Aesar). The electrolyte was stirred with a magnetic bar during the anodization. Finally, the films were cleaned in an ultrasonic bath during 5 min in MilliQ water, left to dry in air and heat treated at $450°C$ ($10°Cmin^{-1}$) during 30 min, to obtain Anatase polymorph (21).

SEM images were acquired with a high field emission microscopy JSM-7600F, with an accelerating voltage of 10.0 kV. The morphologic characteristics of the formed films were estimated using the software iTEM from Olympus soft imaging solutions.

The energy band-gap (E_g) of the heat treated films was calculated using the Kubelka-Munk method from the absorption spectra obtained with a Varian Cary-100 spectrometer equipped with an integration sphere.

The (photo)electrochemical tests were carried out in a conventional three-electrodes cell equipped with a quartz window to allow the UV light illumination to the entire portion (1.23 cm^2) of the TiO_2 nanotube film exposed to the electrolyte. An Ag/AgCl (3M KCl) electrode was employed as reference electrode and a graphite bar (99.999% Alfa Aesar) as counter-electrode. The 0.1 M $HClO_4$ aqueous electrolyte used for films characterization was prepared employing milli-Q water (18.2 $M\Omega cm^{-1}$) and $HClO_4$ (JT Baker, 69%). Before each test the electrolyte was bubbled with N_2 gas during 30 min. The illumination was made using a Newport Q Housing (Model 60025) equipped with a

Hg arc lamp of 100 W. The (photo)electrochemical characterizations were carried out with a BAS-Epsillon potentiostat coupled to a personal computer and, EIS measurements were carried out in an EG&G PAR model 283 potentiostat/galvanostat, coupled to a SI model 1260 Solartron frequency response analyzer.

Results and Discussions

Morphologic characterization

The effect of electrolyte and the formation voltage (V_F) over the morphology of the films was characterized by SEM images. SEM images (top and transversal views) for the potentiostatically growth TiO_2 nanotube films, heat treated at 450°C are shown in Figure 1. Highly ordered TiO_2 nanotube films were obtained from a formation voltage of 10 V, for both electrolytes here employed (NH_4F and HF). Increasing the V_F led to an enlargement in the tube diameter in both electrolytes; however, at a formation voltage of 25 V the nanotube structure collapse resulting in a sponge-like structure, probably due to the more significant chemically attack under this condition. Previous work devoted to the study of the TiO_2 nanotubes growth in aqueous electrolyte has shown that between 25 and 30 V is found a limiting voltage to grow this type of structure in aqueous electrolytes (22).

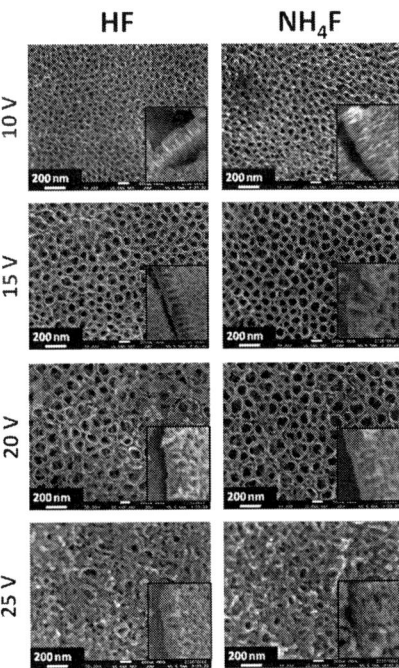

Figure 1. Influence of fluoride counter-ion and formation voltage on the morphology of films formed by anodization during 1 h in 0.1 H_2SO_4/0.05 M F^- electrolytes and heat treated at 450 °C. The transversal view of the film is inserted in each image.

The morphological parameters derived from the SEM images employing the software iTEM from Olympus soft imaging solutions are listed in the Table 1. It is worth mentioning that each data reported in the figure is an average of 100 measurements, except for the film thickness that is an average of 25 measurements. The wall thickness and the tube to tube distance seem to be very similar for the films obtained in both HF and NH₄F. Meanwhile, the tube diameter and the film thickness are considerably changed by the nature of the fluoride counter-ion employed in the anodization electrolyte. On the one hand, films with higher tube diameter in the NH₄F electrolyte were obtained; and on the other hand, the film thickness noticeably shrinks when NH₄F was employed in the anodization bath. Both measurements indicate that titanium oxide grown in NH₄F acid electrolyte is more prone to chemical attack than those oxides grown in HF. In a previous work carried by the authors (20), it was shown by $in\text{-}situ$ EIS study of the anodic oxide film growth, that the presence of NH_4^+ ions in the electrolyte favors the F^- ions insertion in the lattice of the anodic oxide, making it more prone to the chemical attack. This behavior was attributed to the NH_4^+ adsorption on the oxide/electrolyte interface, increasing the F^- ion adsorption on the TiO_2 surface, and fostering its insertion in the oxide lattice.

TABLE I. Morphological parameter of the TiO_2 nanotubes formed, estimated from SEM images in Figure 1.

V_F (V)	Wall thickness (nm)		Tube to tube distance (nm)		Tube diameter (nm)		Film thickness (nm)	
	HF	NH₄F	HF	NH₄F	HF	NH₄F	HF	NH₄F
10	10.2	8.9	57.8	58.5	34.9	47.7	150.8	113.4
15	11.2	11.6	94.7	95.6	54.1	71.3	232.6	177.0
20	12.2	13.4	127.1	127.9	93.7	93.8	253.0	204.1
25	--	--	--	--	--	--	159.1	237.3

UV-Vis characterization

The band gap (E_g) of the formed films at a V_F of 15 V in both electrolytes and heat treated at 450 °C, was evaluated from its absorption spectra using the Kubelka-Munk method for indirect allowed transitions (23) (refer to Figure 2). The E_g values obtained were of 3.3 eV and 3.5 eV for the films grown in HF and NH₄F, respectively. These values are slightly higher than commonly reported E_g for the TiO_2 anatase phase (3.2 eV), probably due to the morphology of the TiO_2 films here prepared. The differences observed in the E_g measured for the films indicate that electrolyte composition may alter the electronic structure of the finally obtained oxide, even after a heat treatment. Particularly, the film growth in NH₄F electrolyte shows a more marked electronic transition, than film growth in HF, sustaining the assumption of an alteration in the properties of the formed oxide.

Semiconductor properties

The semiconductor properties of the formed films were estimated using the Mott-Schottky equation. For this purpose, the EIS spectra was measured from the formed films in 21 different potentials within the potential window ($-0.5 \leq E \leq 0.5$ V vs Ag/AgCl (3M/KCl)) in which the films do not show any faradaic currents in 0.1 M $HClO_4$ (results not shown here). The EIS spectra experimentally obtained were fitted with an electric equivalent circuit reported in the literature for porous films (13), and the space charge capacitance was derived from it. The values of the flat band potential (E_{fb}) and donor density (N_d) are listed in the Table 1, for all the films growth in this study.

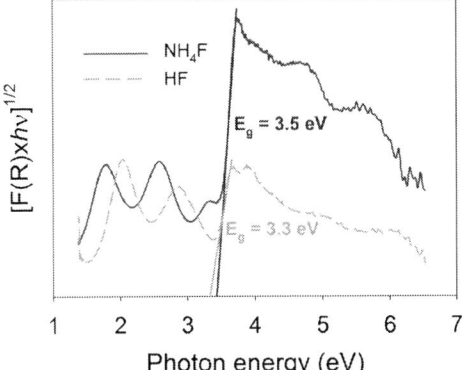

Figure 2. Graphical representation of modified Kubelka–Munk function for indirect allowed transitions (23), for the TiO$_2$ nanotubes films formed at 15 V during 1 h in 0.1 M H$_2$SO$_4$/0.5 M F⁻ and heat treated at 450 °C.

TABLE II. Semiconductor properties of the TiO$_2$ nanotubes formed and heat treated at 450 °C.

V_F (V)	Flat band potential, E_{fb} (mV) vs Ag/AgCl (3M/KCl)		Donor density, $N_d \times 10^{-22}$ (cm^{-3})	
	HF	NH$_4$F	HF	NH$_4$F
10	220.7	112.6	2.0	9.0
15	28.7	-44.6	19.4	75.0
20	61.1	-44.3	39.7	78.7
25	219.0	37.4	30.3	54.1

The E_{fb} values show the same tendency with V_F indistinctly the electrolyte employed during its anodization, exhibiting a minimum for a V_F of 15 V. However, when NH$_4$F was employed instead HF, E_{fb} takes lower values even reaching negative values. On the other hand, the N_d values increase with the V_F reaching a maximum at 20 V, and then decrease for a V_F of 25 V. The alteration of the tendency at 25 V might be associated to the change in the film morphology that changed from nanotubes to a sponge-like film (Figure 1). Additionally, the N_d were also altered by the electrolyte used for the film growth, reaching higher values when NH$_4$F was employed instead HF. The dependence of semiconductor properties on the electrolyte employed during its growth, can be accounted for the fact that the presence of NH$_4^+$ ions in the anodization bath favors the insertion of fluoride ions in the lattice of TiO$_2$ films (20). When the as-formed TiO$_2$ nanotube film is heat treated at 450°C, not only does the amorphous structure of the oxide crystallize to anatase, it also eliminate the F⁻ ions present in the material (10). However, the presence of a higher quantity of fluoride ions inside the material may induce a major crystallization in the film, or the formation of new defects altering the semiconductor properties of the formed films. This can be reflected as an increase in the density of donor in the material (*n-type* defects) and lower values of E_{fb}.

Photoelectrochemical characterization

Open Circuit Potential (OCP). When an *n-type* semiconductor electrode immersed in an aqueous electrolyte is illuminated, in the absence of an oxidant agent, the photoexcited electrons are accumulated in the conduction band of the semiconductor, or the energetic states near it; while the holes in the valence band are rapidly transported to the oxide/electrolyte interface to carry the oxidation of a reductant specie in the solution (in this case the water) (17, 24). These phenomena cause a change in the open circuit potential of the semiconductor material towards more negative values, representing an indirect measurement of the electrons energy level in the semiconductor under illumination. However, it is worth mentioning, that the values obtained with these measurements must be carefully handled, since the open circuit potential under illumination depends on the intensity and electromagnetic spectra of the light employed for this purpose (25).

In the Figure 3 are shown the OCP measurements in the dark and under illumination for the films prepared at different V_F, in HF electrolyte (Figure 3 (a)) and NH$_4$F electrolyte (Figure 3 (b)). When the TiO$_2$ porous films were illuminated a rapidly variation of the open circuit potential (up to ~500 mV) was observed, followed by a relaxation stage; then the illumination was interrupted during 3 minutes, and the film was illuminated for a second time, showing a similar rapidly variation of the OCP, but in this time, an almost stationary value was reached for all the films.

The behavior observed during the first illumination step could be due to the filling of the energetic states near the conduction band of the TiO$_2$, and then, when the film was again illuminated, this stage was not shown because these energetic states are already occupied by the photoexcited electrons during the first perturbation (19).

The open circuit potential under the second illumination presents a similar dependence with V_F as the E_{fb} in the Table 2; confirming that when NH$_4$F is employed in the anodization bath, there is a variation of the energetic states in the valence band towards more negative potentials, probably due to the higher insertion of F$^-$ in the TiO$_2$ lattice (20).

Chronoamperometry. The linear voltammetry evaluation (results not shown) of the formed films did not show the typical saturation current obtained during the photoelectrochemical characterization; however, a potential of 1.25 V was chosen for the evaluation of the photocurrent generated under illumination by chronoamperometric pulses, since the photocurrent just showed a slightly variation at higher potentials. During this analysis, a potential of 1.25 V was imposed in the dark and after two minutes, the films were illuminated during a period of one minute, to observe the photocurrents generated due to the water oxidation. In order to evaluate the stability of the photogenerated currents, three additional perturbations with illumination during a period of time of one minute was carried, after letting the system rest in the dark for a period of one minute.

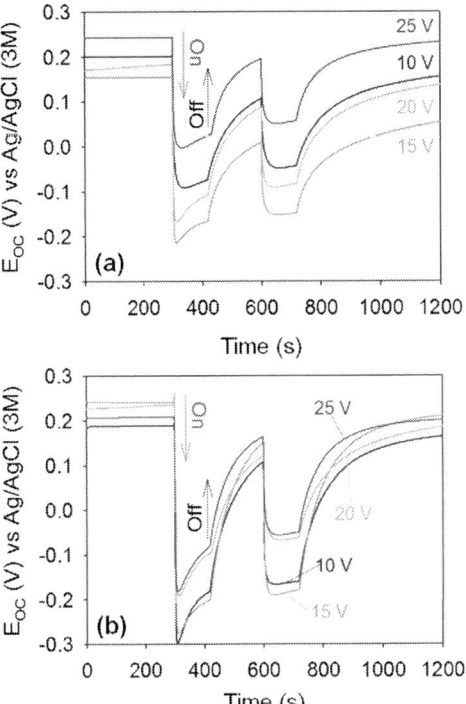

Figure 3. Effect of the UV Light illumination over the open circuit potential versus time curves, measured in a 0.1 M HClO₄ solution, for TiO₂ nanotube films (heat treated at 450 °C) previously prepared by anodization under different formation voltage (indicated in the figure), in 0.1 M H₂SO₄ and: (a) 0.05 M HF or (b) 0.05 M NH₄F.

In the Figure 4 are shown the currents registered in the dark and under illumination for the films formed under different V_F in a HF electrolyte (Figure 4 (a)) and in a NH₄F electrolyte (Figure 4 (b)). In both cases a maximum photocurrent was registered for the films formed at a V_F of 15 V after which the recorded photocurrents decreased.

It is worth mentioning that the film growth in HF at a V_F of 25 V not only showed a decrease in the photogenerated currents, but also a detriment in the stability of the photogenerated current as the illumination period was repeated. These considerable shrinking of the photocurrent and stability of the film might be due to the drastic morphological change observed under this V_F (Figure 1 and Table I).

The photoelectrochemical performance of the potentiostatically growth films seem to be controlled by the extend in the modification of the semiconductor properties, induced by the electrolyte employed for its growth in aqueous electrolytes, showing higher photocurrent for the films with a greater density of donor and a lower flat band potential. This knowledge can be helpful for obtaining TiO₂ nanotubes films with paramount performance for water oxidation, showing that not only operational variable as potential (or potential program) and anodization time must be optimized, but also bath composition is needed to be resolved in order to improve the photoelectrochemical performance of the film finally formed.

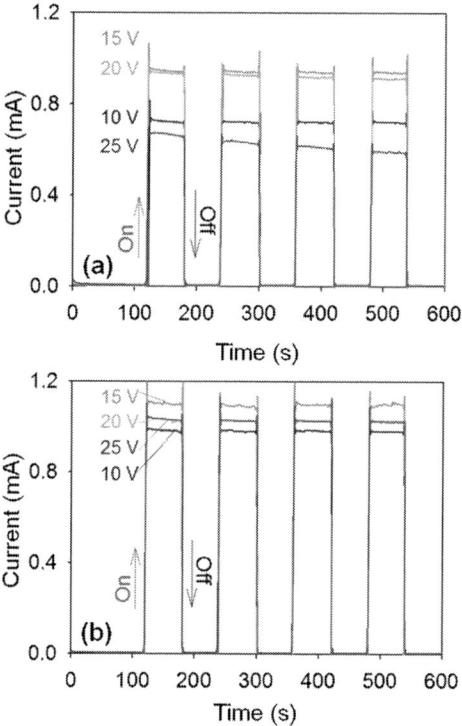

Figure 4. Effect of the UV Light illumination over the current versus time curves, measured imposing a potential of 1.25 V vs Ag/AgCl in a 0.1 M HClO$_4$ solution, for TiO$_2$ nanotube films (heat treated at 450 °C) previously prepared by anodization under different formation voltage (indicated in the figure), in 0.1 M H$_2$SO$_4$ and: (a) 0.05 M HF or (b) 0.05 M NH$_4$F.

Conclusions

SEM images of Ti anodic films showed that the use of NH$_4$F instead of HF provokes a larger and thinner tube diameter, indicating a more significant chemical attack of the anodic oxide formed in this electrolyte. Additionally, the morphology of the obtained TiO$_2$ films drastically changed at a formation voltage of 25 V, passing from a nanotube vertically ordered structure to sponge-like film. The band-gap measurements evaluated from absorption spectra, and the semiconductor properties estimated from Mott-Schottky equation, showed that films formed in NH$_4$F electrolytes present different properties than films formed in HF; exhibiting a greater band-gap, a higher density of donor and lower flat band potential. These variations certainly impacted the photoelectrochemical performance of the formed films showing the generation of more negative photopotential at open circuit, and the higher photocurrents at a potential of 1.25 V vs Ag/AgCl (3 M KCl). Finally, it is worth mentioning, that for both electrolytes (HF and NH$_4$F) it was found a better photoelectrochemical performance for the films growth at a formation voltage of 15 V, that could be attributed to their highly ordered TiO_2 nanotube structure and semiconductor properties (lower flat band potential and high density of donor).

Acknowledgments

This work has been given the financial support from CONACyT (Project CB-2008/105655). Próspero Acevedo Peña is grateful to CONACyT for the PhD grant through the program *doctorados nacionales*. The authors thank to Dra. Patricia Castillo from *Laboratorio Central de Microscopía Electrónica* (UAM-I) for her assistance in SEM images.

References

1. Y. Lai, L. Sun, Y. Chen, H. Zhuang, C. Lin and J.W. Chin, *J. Electrochem. Soc.*, **153**, D213 (2006).
2. J. Wang and Z. Lin, *Chem-Asian J.*, **7**, 2754 (2012).
3. V.K. Mahajan, M. Misra, K.S. Raja and S.K. Mohapatra, *J. Phys. C*, **41**, 125307 (2008).
4. J. Li, H. Yu and C-J. Lin, *J. Electrochem. Soc.*, **154**, C631 (2007).
5. X. Wu, Y. Ling, L. Liu and Z. Huang, *J. Electrochem. Soc.*, **156**, K65 (2009).
6. N.K. Allam, K. Shankar and C.A. Grimes, *J. Mater. Chem.*, **18**, 2341 (2008).
7. L.X. Sang, Z.Y. Zhang and C.F. Ma, *Int. J. Hydrogen Energy*, **36**, 4732 (2011).
8. S. Liang, J. He, Z. Sun, Q. Liu, Y. Jiang, H. Cheng, B. He, Z. xie and S. Wei, *J. Phys. Chem. C*, **116**, 9049 (2012).
9. D-S. Kong, X-D. Zhang, j. Wang, C. Wang, X. Zhao, Y-Y. Feng and W-J. Li, *J. Solid State Electrochem.*, **17**, 69 (2013).
10. P. Roy, S. Berger and P. Schmuki, *Angew. Chem.. Int. Ed.*, **50**, 2904 (2011).
11. B. Liu, K. Nakata, S. Liu, M. Sakai, T. Ochiai, T. Muramaki, K. Takagi and A. Fujishima, *J. Phys. Chem. C*, **116**, 7471 (2012).
12. A.G. Muñoz, Q. Chen and P. Schmuki, *J. Solid State Electrochem.*, **11**, 1077 (2007).
13. S.A.A. Yahia, L. Hamadou, Z. Kadri, N. Benbrahim and E.M.M. Sutter, *J. Electrochem. Soc.*, **159**, K83 (2012).
14. P. Pu, H. Cachet, N. Laidani and E.M.M. Sutter, *J. Phys. Chem. C*, **116**, 22139 (2012).
15. M. Radecka, M. Rekas, A. Tenczek-Zajac and K. Zakrzewska, *J. Power Sources*, **181**, 46 (2008).
16. P. Acevedo-Peña, G. Vázquez, D. Laverde, J.E. Pedraza-Rosas, J. Mariquez and I. González, *J. Electrochem. Soc.*, **156**, C377 (2009).
17. P. Acevedo-Peña, J. Manríquez and I. González, *ECS Trans.*, **29**, 183 (2010).
18. P. Acevedo-Peña and I. González, *ECS Trans.*, **36**, 257 (2011).
19. P. Acevedo-Peña and I. González, *J. Electrochem. Soc.*, **159**, C101 (2012).
20. P. Acevedo-Peña, D. Valdez and I. González, *J. Electrochem. Soc.*, submitted.
21. Y-K. Lai, J-Y. Huang, H-F. Zhang, V-P. Subramaniam, Y-X. Tang, D-G. Gong, L. Sundar, L. Sun, Z. Chen and C-J. Lin, *J. Hazard. Mater.*, **184**, 855 (2010).
22. S. Mahshid, A. Dolati, M. Goodarzi, M. Askari and A. Ghahramaninezhad, *ECS Trans.*, **28**, 67 (2010).
23. R. López and R. Gómez, *J. Sol-Gel Sci. Technol.*, **7**, 611 (2012).

24. P. Acevedo-Peña and I. González, *J. Solid State Electrochem.*, **17**, 519 (2013).
25. R. Beranek, *Adv. Phys. Chem.*, **20**, 786759 (2011).

CHAPTER 4

ELECTRODEPOSITION AND
SEMICONDUCTOR METALLIZATION

Evidence of phosphazene steps formation on InP by cyclic voltammetry studies and XPS analyses in liquid ammonia (-55°C)

C. Njel, A.M. Gonçalves*, D. Aureau, D. Mercier, A. Etcheberry

University of Versailles St Quentin-en-Yvelines. Institut Lavoisier de Versailles
UMR- CNRS 8180
45, Avenue des Etats-Unis 78000 Versailles - France.

In liquid ammonia (-55°C), a stable phosphazene like film has been already revealed on InP. This work reports, for the first time, the evidence of different stages of this film on InP in this relevant non aqueous solvent. These results were carried out by a fully coupled approach with cyclic voltammetry, capacitance measurements and XPS analyses.

Introduction

Indium phosphide is an attractive material for high speed optoelectronic devices. Its integration in electronic devices is however still complicated due to the anarchic stoichiometry of oxides InP surface. Several attempts to decrease the density of traps at the insulator/InP interfaces have already been made using wet process, for instance inorganic sulfite chemistry (1, 2). However, these passivation treatments were only efficient for a limited period of use. A degradation of the surface network was indeed rapidly detected. In past decades, in order to avoid oxide formation at the InP/electrolyte interface, many electrochemical processes have been studied in non aqueous solvents but the interface was still ruled by moisture, i.e. water chemistry (see e. g. (3, 4)).
As a singular electrolyte, acidic liquid ammonia (NH_3 *liq*) is distinguished from the other non aqueous solvents since moisture does not interact at the interface (5). Significant prospects of this uncommon solvent have been recently reported through unusual electrochemical mechanisms onto III-V semiconductors. For high anodic current densities (> 2 $mA.cm^{-2}$) and high interfacial potential (> 2 V/Silver Reference Electrode, noted SRE), both original porosification of InP surface and ammonia oxidation are evidenced (6, 7). For low anodic current density, in liquid ammonia, the electrochemical formation of a monolayer protective film on InP was recently published (8, 9).

This passivation treatment was performed by cyclic voltammetry under anodic overvoltage either in the dark or under illumination (10, 11). Anodic photo-galvanostatic process was also observed using controlled current techniques from 1 to 10 $\mu A.cm^{-2}$. Whatever the method used, a phosphazene like film was revealed by X-ray Phoelectron Spectroscopy (XPS). The perfect coverage of the InP matrix by this nitrogenated film is evidenced by this fully coupled approach using electrochemistry and XPS analyses. In open air, no degradation of the thin film was evidenced by XPS, photo-luminescence and electrochemistry (11). The high chemical stability of this nitrogenated film opens a new way for III-V semiconductors integration in electronic devices.

In this paper, we report the different stages of the phosphazene like film formation on *n*-InP by using *in-situ* electrochemical techniques such as cyclic voltammetry and capacitance measurements. For each stage, the sample is removed from the electrochemical cell and the chemical evolution of *n*-InP surface is analyzed by XPS. This coupled approach provides the successive and resulting chemical evolution of the interface during the anodic treatment in NH_3 *Liq*.

Experimental

InP wafers *(-n and –p)* with a (100) orientation and a doping density of $10^{18}cm^{-3}$ (InPact Electronic Materials, Ltd) are cleaved in $0.5\times0.5cm^2$ square pieces. Prior to the passivation treatment, samples are chemomecanically polished in a Br_2-CH_3OH solution (2%), fully rinsed in high purity CH_3OH and dried under a N_2 flux. Residual oxides on the surface are eliminated by a short immersion in a 2M HCl solution.

Liquid ammonia (NH_3 *Liq*) is condensed from gaseous ammonia ("electronic grade" from Air Liquide). The volume of liquid ammonia is $150\ cm^3$ and is maintained at -55°C in a cryostat. High purity salts of NH_4Br ($10^{-1}M$, Sigma Aldrich) are added in NH_3 *liq* to provide the electric conductivity of the electrolyte and to maintain the pH at 1 (referred to ammonia scale). The electrochemical set-up is a classical 3 electrode device connected to a Parstat 2273 potensiostat. All potentials are measured *vs.* a silver reference electrode (SRE) (12). A large smooth platinum electrode is also used as a counter electrode. The deaeration of the medium is performed under an argon stream. An optical fiber is immerged in the cryostat and connected to a tungsten lamp which allows the illumination of InP through the electrochemical cell. The electrochemical treatment of the InP surfaces is performed by cyclic from -0.5 V to +1.7 V (20mV/s) under illumination to minimize the resultant interfacial potential.

At open circuit, a drastic and controlled transfer procedure is used to avoid thermal shocks. Prior to surface analysis, samples are carefully rinsed in pure NH_3 *Liq* and transferred under controlled atmosphere conditions (N_2). High resolution XPS surface experiments are achieved with a VG 220i-XL Escalab with a monochromatic Al source, 10kV primary energy, 16 mA emission intensity and 20 eV pass energy over $1\times1mm^2$ analysis area. Spectrometer calibration is performed using the manufacturer's procedure, which is completed by a self-consistent check on sputtered copper and gold samples, based on the ASTM E902-94 recommendation.

Results and Discussion

After a suitable chemical etching of InP, an anodic wave is only observed during the first positive scan in acidic NH_3 *liq*. under illumination. Even if this electrochemical phenomenon is only detected during the first scan, this anodic wave is highly reproducible. A charge close to 8 $mC.cm^{-2}$ is involved in this wave. Different positions on the wave, from A to F, are reported in the Fig. 1. For each position on the wave, a corresponding anodic charge is involved (Q_A= 0.07 $mC.cm^{-2}$, Q_B= 0.75 $mC.cm^{-2}$ and Q_C= 1.35 $mC.cm^{-2}$, Q_D= 3.9 $mC.cm^{-2}$, Q_E= 6.6 $mC.cm^{-2}$, Q_F= 8 $mC.cm^{-2}$). As it is reported in the figure 1, during the first scan, the anodic current gradually increases until +1.3 V/SRE. This potential position corresponds to the anodic peak (position C). This anodic wave is promptly followed at +1.7 V SRE by a strong current increase without obvious limitation. This anodic current is associated to the dissolution of InP of the semiconductor (13). As soon as the anodic wave has been recorded, the anodic behavior of InP was drastically modified during the reverse and following scans. The anodic wave did not emerge again, only the unlimited current (from +1.7 V) was still observed.

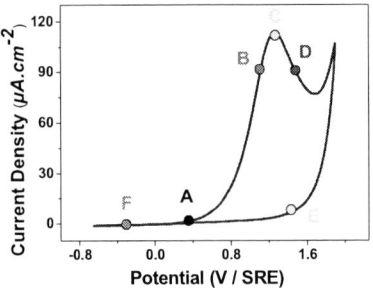

Figure 1. Cyclic voltammogram under illumination in liquid ammonia (T= -55°C, [NH$_4$Br] =0.1 M) observed during the first anodic scan with a scan rate of 20 mV.s^{-1}. Potentials are measured against SRE. Potential stops on the wave from A to F for different anodic charge (Q$_A$= 0.07 mC.cm^{-2}, Q$_B$= 0.75 mC.cm^{-2} and Q$_C$= 1.35 mC.cm^{-2}, Q$_D$= 3.9 mC.cm^{-2}, Q$_E$= 6.6 mC.cm^{-2}, Q$_F$= 8 mC.cm^{-2}).

In order to explore the progress of the film formation on n-InP, in-$situ$ interfacial capacitances are carried out for each position on the anodic wave (A to F). Mott-Schottky plots measurements are indeed reported in the figure 2.

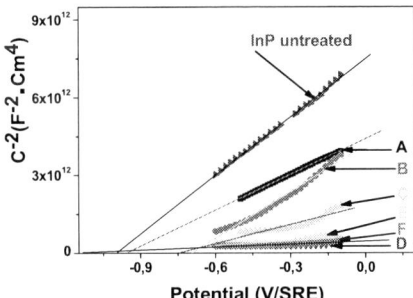

Figure 2. Mott Schottky plots, in the dark, on n-InP, in liquid ammonia (T= -55°C, [NH$_4$Br] =0.1M) with a scan rate of 20 mV.s^{-1}, for a frequency, f=1.03 kHz. Potentials are considered against SRE. Plots are plotted after each stop on the anodic wave from A to F with the corresponding anodic charge: Q$_A$= 0.07 mC.cm^{-2}, Q$_B$= 0.75 mC.cm^{-2} and Q$_C$= 1.35 mC.cm^{-2}, Q$_D$= 3.9 mC.cm^{-2}, Q$_E$= 6.6 mC.cm^{-2}, Q$_F$= 8 mC.cm^{-2}. "InP untreated" corresponds to a non modified surface for which no anodic overvoltage is applied.

All the experiments were performed at a pH of 1 according to the liquid ammonia scale (14). At this pH and on *n*-InP, freshly deoxidized, Mott-Schottky plot (Fig. 3 "InP untreated") gives the expected flat band potential by extrapolation of the straight line (15). However, as soon as the anodic treatment is initiated a drastic modification of the slope plot is observed. As a function of the anodic charge, a regular decrease of the slope is indeed revealed. The more the anodic charge raises the more the Mott-Schottky plot declines. A stabilization of the Mott-Schottky plot is observed for the position F on the anodic wave which corresponds to an anodic charge close to 8 mC.cm^{-2}. The electrical charge distribution at the interface is evidenced by capacitance measurements. The variation of the interfacial capacitance with the applied potential can be then used as an *in situ* probe of the film electrical properties. According the anodic, charge and in the range of explored potential, the space charge layer of InP is progressively not detected. The applied potential at the interface should be then shared with the passivating film and the Helmholtz layer.

In order to connect the increase of the anodic charge with the chemical evolution of the surface, XPS analyses were performed for each position in the anodic wave (A to F). After each position, a sample is removed slowly from the electrochemical cell to avoid thermal shocks. The sample is rinsed by a long dumping in purest liquid ammonia (without conducting salt). After the rinsing step, the semiconductor is removed toward XPS analyzer using a transfer procedure avoiding any air contamination. The anodic linear sweep is stopped at each point (A to F). There is then one new sample by potential stop. Each position in the wave atomic surface ratios are indeed reported in the Fig. 3.

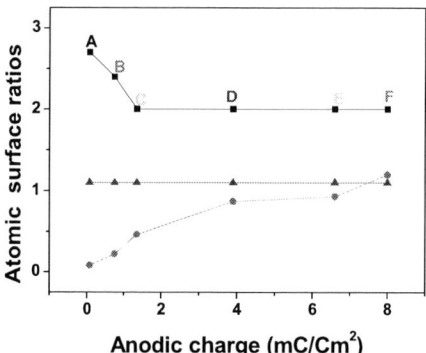

Figure 3. Atomic surface rations as a function of the anodic charge after each potential stops on cyclic voltammetry. The linear sweep was stopped at each point and then each sample is analyzed. (Q_A= 0.07 mC.cm^{-2}, Q_B= 0.75 mC.cm^{-2} and Q_C= 1.35 mC.cm^{-2}, Q_D= 3.9 mC.cm^{-2}, Q_E= 6.6 mC.cm^{-2}, Q_F= 8 mC.cm^{-2}).
-■- :N_{1s}/P_{2p}^{HE} ; -●- : $P_{2p}^{HE}/P_{2p}^{matrix}$; -▲- : $In_{3d}^{matrix}/P_{2p}^{matrix}$

In spite of the gradual anodic charge applied at the interface by cyclic voltammetry, no chemical deviation of indium and phosphorus from the matrix is observed. The atomic ratio of $In_{3d}^{matrix}/P_{2p}^{matrix}$ is still equal to 1 whatever the anodic charge. However, as a function of the load, a contribution of phosphorus at high energy (133 eV) is evidenced (8). The atomic ratio $P_{2p}^{HE}/P_{2p}^{matrix}$ gradually increases from 0 to around 1. The beginning of the anodic scan (A and B) is also recorded by a striking contribution of nitrogen. The atomic ratio N_{1s}/P_{2p}^{HE} starts with 2.5 to finally stabilize at 2. As expected, theses atomic ratios correspond to those evidenced for the formation of the phosphazene like film on InP (8). Even if an enhancement of phosphorus at high energy and nitrogen contribution is evidenced, the saturation of both ratios (N_{1s}/P_{2p}^{HE}, $P_{2p}^{HE}/P_{2p}^{matrix}$) is not detected for the same anodic charge. An anodic charge around 1 mC.cm^{-2} is required for the stabilization of the atomic ratio of N_{1s}/P_{2p}^{HE} whereas an anodic charge close to 8 mC.cm^{-2} is involved for the stabilization of the atomic ratio of $P_{2p}^{HE}/P_{2p}^{matrix}$. Only the evolution of the last ratio ($P_{2p}^{HE}/P_{2p}^{matrix}$) is consistent with the progress of the capacitance measurements according the anodic charge.

Conclusions

For the first time, the evidence of different stages of phosphazene like film on InP in liquid ammonia (-55°C) are investigated using a fully coupled approach with *in-situ* electrochemistry (cyclic voltammetry, capacitance measurements) and XPS analyses. As the anodic wave is described, a drastic evolution of the Mott-Schottky plot is observed. A strong decrease of the Mott-Schottky plot is reported. This behavior can be used as an *in situ* probe of the electrical properties of the film. The Mott-Schottky plot evolution agrees with the strong chemical evolution of the surface. A charge close to 8 mC.cm^{-2} is indeed required for the stabilization of the atomic ratio $P_{2p}^{HE}/P_{2p}^{matrix}$. Only the stabilization of the atomic ratio N_{1s}/P_{2p}^{HE} requires a deeply impacted lower anodic charge (around 1 mC.cm^{-2}) whereas no stabilization of Mott-Schottky plot is observed. The difference of charge required for the stabilization of both ratios can suggest a mechanism which involves multi step (electrochemical or/and chemical) for the formation of the phosphazene like film on *n*-InP.

References

1. Y. H. Jeong, S. K. Jo, B. H.Lee, T.Sugano, *IEEE Electron Device Letters.*, **16**, 109 (1995).
2. D. N.Bose, Y.Ramprakash, S. Basu, *Mater. Lett.*, **88**, 364 (1989).
3. T. A. Abshere and G. L. Richmond, *J. Phys. Chem. B.*, **104**, 1602 (2000).
4. T. A. Abshere, G. L. Richmond, *J. Phys. Chem. B.*, **103**, 7911 (1999).
5. A-M. Gonçalves, C. Mathieu, M. Herlem and A. Etcheberry, *J. Electroanal. Chem.*, **462**, 88 (1999).
6. A-M. Goncalves, L. Santinacci, C. David, C. Mathieu, M. Herlem, A. Etcheberry, *Phys. Status. Solidi A.*, **204**, 1286 (2007).
7. O. Seitz, C. Mathieu, A.-M Gonçalves, M. Herlem, A. Etcheberry, *J. Electrochem. Soc.*, **150**, E461 (2003).].
8. A.-M. Gonçalves, N. Mézailles, C. Mathieu, P. Le Floch, A. Etcheberry, *Chem. Mat.*, **22**, 3114 (2010).
9. A.-M. Goncalves, C. Mathieu, N. Mézailles, A. Etcheberry, *ECS Transactions 8, Processes at the Semiconductor-Solution Interface 4 Electrochem. Soc. Electrochem. Soc.*, **1101**, 1326 (2011).
10. A.-M. Gonçalves, O. Seitz, C. Mathieu, M. Herlem, A. Etcheberry, *Electrochem. Comm.*, Vol **10/2**, 225 (2007).
11. A.-M. Goncalves, C. Mathieu, A. Etcheberry, *J. Electrochem. Soc.*, **159**, (2) C97 (2012).
12. D. Guyomard, M. Herlem, C. Mathieu, J. L. Sculfort, J. *Electroanal. Chem. Interfacial Electrochem.*, **216**, 101 (1987).
13. A.-M. Gonçalves , C. Mathieu, M. Herlem, A. Etcheberry, *Electrochimica Acta*, **46**, 2835 (2001).
14. J. Jander, Anorganische und allgemeine Chemie in flüssigen Ammoniak. **Part I**, Friedr. (Eds. Vieweg & Sohn), Braunschweig (1966).
15. A-M. Gonçalves, C. Mathieu, M. Herlem, A. Etcheberry, *Electrochimica Acta* **55**, 7413 (2010).

Electroless Metallization of Silicon Using Metal Nanoparticles as Catalysts and Binding-Points

S. Yae[a], M. Enomoto[a], H. Atsushiba[a], A. Hasegawa[a], C. Okayama[a], N. Fukumuro[a], S. Sakamoto[a, b], and H. Matsuda[a]

[a] Department of Materials Science and Chemistry, Graduate School of Engineering, University of Hyogo, 2167 Shosha, Himeji, Hyogo 671-2280, Japan
[b] Nippon Oikos Co., Ltd., 3-6-16 Mitsusadadai, Yahatanishi, Kitakyushu, Fukuoka 807-0805, Japan

Gold nanoparticles on silicon work not only as catalysts to initiate autocatalytic electroless metal deposition but also as binding-points between the deposited metal film and the silicon surface. Conventional catalysts of autocatalytic electroless metal deposition, such as palladium and silver, require heat treatments or anchor formation to obtain practical adhesion of deposited metal films on silicon. Gold nanoparticles can directly produce adhesive metal films on flat silicon surfaces without any treatments. Cross-sectional transmission electron microscopic observation reveals that gold nanoparticles form an alloy with silicon at the room temperature. This alloy is expected to improve the adhesion of metal film on silicon.

Introduction

The metallization of silicon surfaces is widely used from nanometer scale for ULSI devices to decimeter scale for solar cells. Adhesive metal-film formation on silicon is important to metalize silicon surfaces for obtaining infallible electrical contacts in various devices. Autocatalytic electroless deposition, which is a conventional method for metal-film formation on such nonmetallic substrates as ceramics, glass, and semiconductors (1, 2), requires the surface activation (catalyzation pretreatment) of substrates, generally using tin and palladium (1-3). For silicon substrates, obtaining adhesive metal films with conventional catalyzation pretreatments is difficult. Heat treatments before and after deposition improved the adhesion of electrolessly deposited metal films on silicon (4, 5). We recently developed a new surface-activation process for the direct electroless deposition of adhesive metal films on silicon substrates that consists of three steps (6, 7): (step 1) metal nanoparticle formation by electroless displacement deposition expressed by Eqs. [1] and [2] (8); (step 2) silicon nanopore formation by metal-assisted hydrofluoric acid etching (9); and (step 3) metal filling in nanopores, that is, metal nanorod formation and metal-film formation on the whole silicon surface by autocatalytic electroless deposition. This method can deposit adhesive metal films on silicon without a heat treatment. In this study, we found that gold nanoparticles have an excellent ability to bind metal films to silicon surfaces and require no nanopore formation of step 2 or a heat-treatment.

$$M^{n+} \rightarrow M + nh^+ \quad (h^+: \text{hole}) \qquad [1]$$

$$Si + 6F^- + 4h^+ \rightarrow SiF_6^{2-} \qquad [2\text{-}1]$$
$$\text{and/or} \quad Si + 4HF_2^- + 2h^+ \rightarrow SiF_6^{2-} + 2HF + H_2 \qquad [2\text{-}2]$$

Experimental

Single-crystalline p-type silicon wafers (CZ, (100), ca. 1 Ω cm, ca. 0.7-mm-thick, Yamanaka Semiconductor) were cut into 2 x 3 cm pieces and used as substrates. The silicon substrates were washed with acetone and etched with a CP-4A solution and then with a 7.3 M (M: mol dm^{-3}) hydrofluoric acid solution. Gold nanoparticles were produced on the silicon substrates by electroless displacement deposition using a 0.5 mM tetrachloroauric (III) acid solution containing 0.15 M hydrofluoric acid for 120 s at a solution temperature of 278 K. Nickel-phosphorus alloy films were formed on the silicon substrates by autocatalytic electroless deposition using a mixture solution of 0.1 M nickel sulfate, 0.3 M sodium phosphinate, 0.1 M succinic acid, and 0.1 M DL-malic acid at a pH of 4.8 adjusted with a sodium hydroxide solution and a solution temperature of 343 K. The thickness of the deposited nickel films was measured by a gravimetric procedure. The adhesion of the electrolessly deposited metal films on the Si substrates was examined by a tape test based on Japanese Industrial Standard JISH8504 that corresponds to ISO 2819 (10). The film was notched at 2 x 2 mm on the Si substrate with a steel blade cutter. An adhesive tape was stuck on the film and peeled by pulling one of its ends under a constant velocity of 2 mm s^{-1}. We used an adhesive tape (Nichiban, CT-18) prescribed for conventional tape tests by JIS that can measure peel energy (10, 11) up to 400 J m^{-2}. The adhesion was evaluated by the metal film area on the Si substrate that remained after the tape test. Surface and cross-sectional inspections of the specimens were performed with a scanning electron microscope (SEM, JEOL JSM-7001F) and a transmission electron microscope (TEM, JEOL JEM-2100).

Results and Discussion

Bright and uniform nickel films were electrolessly deposited on the gold-nanoparticle-modified silicon wafers. Figure 1 shows a 4-inch silicon wafer after gold nanoparticle formation on half of its surface and then immersion of the whole surface in an autocatalytic electroless nickel deposition solution. No metal deposition occurred on the non-gold-nanoparticle-modified area of the silicon wafer. The whole silicon surface of the gold-nanoparticle-modified area was uniformly covered with a mirrorlike metallic color film. No peeling of the metal film occurred by the tape test. Figure 2 shows the percentage of the area of the electrolessly deposited nickel films that remained on the silicon substrates after the tape test as a function of the thickness of the deposited nickel films. No peeling occurred of the metal film thinner than 1.3 μm. This result indicates that whole films, whose thickness is less than 1.3 μm, have higher adhesive energy than 400 J m^{-2}. The remaining area, which is the adhesion area of the electrolessly deposited nickel films on silicon, decreased as the film thickness exceeded 1.3 μm. We produced a metal pattern on silicon using this method. A photoresist pattern was formed on the silicon surface. Gold nanoparticles were formed on the bare parts of the silicon surface by electroless displacement deposition. The resist was removed with a solvent. Then we performed autocatalytic electroless nickel deposition by immersing the silicon wafer into a plating solution. As shown in Fig. 3, a sharp and clear metal pattern was formed on the silicon surface. This metal pattern was evaluated highly adhesive by the tape test.

Figure 1. 4-inch p-type silicon wafer after gold nanoparticle formation on half of its surface and then immersion of the whole surface in an autocatalytic electroless nickel deposition solution. Gold-nanoparticle-modified area is on the left half of the wafer, mirroring the University of Hyogo emblem as a bare surface (right half) of a mirror polished wafer.

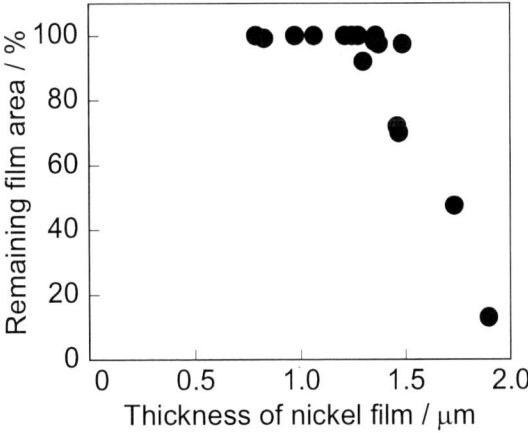

Figure 2. Percentage of nickel film area that remained on silicon substrates after tape test as a function of film thickness.

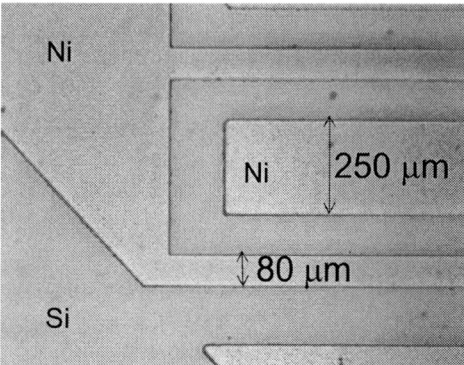

Figure 3. Optical microscopic image of electrolessly deposited nickel pattern on silicon.

Figure 4 shows the surface SEM and cross-sectional TEM images of the silicon after the gold nanoparticle formation. Spherical nanoparticles, 6–25 nm in diameter and 3.5×10^{11} cm^{-2} in particle density, were scattered with a few tens of nm distance between each one. The cross-sectional image indicates that gold atoms diffused into the silicon and formed a gold-silicon alloy. No such alloy was found in the silver nanoparticle case that requires nanoanchor formation to deposit adhesive metal film on the silicon substrates. A silicon and gold metallic alloy was formed at the interface between silicon and gold even at room temperature, and silicon easily diffused in gold and formed a silicon oxide layer on the surface of gold under the presence of oxygen gas (12). This alloy formation is expected to improve the adhesion of metal film on silicon. Thus, gold nanoparticles work not only as catalysts to initiate autocatalytic electroless metal deposition but also as binding-points between the deposited metal film and the silicon surface.

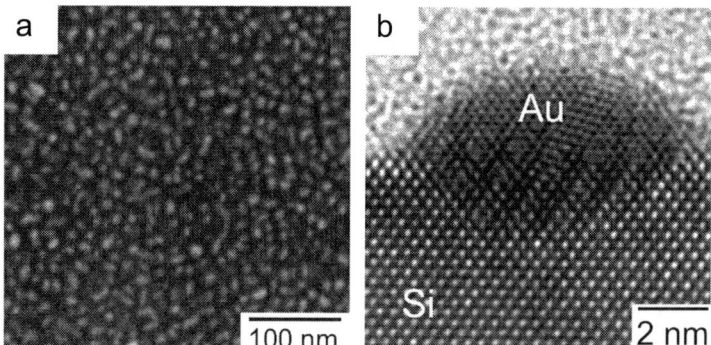

Figure 4. SEM (a: plane-view) and TEM (b: cross-section) images of gold-nanoparticle-modified silicon wafer.

Acknowledgments

The present work was partly supported by Grant-in-Aid for Scientific Research (C) (23560875) from JSPS (Japan Society for the Promotion of Science), and FS stage of A-STEP (Exploratory Research of Adaptable and Seamless Technology Transfer Program through Target-driven R&D) from JST (Japan Science and Technology Agency). The authors are grateful to Prof. A. Yamamoto, Department of Materials Science and Chemistry, Graduate School of Engineering, University of Hyogo for the ion-milling of the cross-sectional TEM specimens.

References

1. M. Paunovic and M. Schlesinger, *Fundamentals of Electrochemical Deposition 2nd. Ed.*, p. 153, Wiley Interscience, New York (2006).
2. M. Schlesinger, in *Modern Electroplating 4th. Ed.*, M. Schlesinger and M. Paunovic, Editors, p. 613, Wiley Interscience, New York (2000).
3. K. Yamagishi, N. Okamoto, N. Mitsumata, N. Fukumuro, S. Yae, and H. Matsuda, *Trans. Inst. Met. Fin.*, **82**, 114 (2004).
4. V. M. Dubin, *J. Electrochem. Soc.*, **139**, 1289 (1992).
5. S. Karmalkar and V. P. Kumar, *J. Electrochem. Soc.*, **151**, C554 (2004).
6. S. Yae, K. Sakabe, N. Fukumuro, S. Sakamoto, and H. Matsuda, *J. Electrochem. Soc.*, **158**, D573 (2011).
7. S. Yae, K. Sakabe, T. Hirano, N. Fukumuro, and H. Matsuda, *Phys. Stat. Sol. (c)*, **8**, 1769 (2011).
8. S. Yae, N. Nasu, K. Matsumoto, T. Hagihara, N. Fukumuro, and H. Matsuda, *Electrochim. Acta*, **53**, 35 (2007).
9. S. Yae, Y. Kawamoto, H. Tanaka, N. Fukumuro, and H. Matsuda, *Electrochem. Comm.*, **5**, 632 (2003).
10. K. Ito, N. Fukumuro, S. Yae, and H. Matsuda, *J. Jpn. Inst. Electron. Packag.*, **12**, 130 (2009).
11. J. W. Severin, R. Hokke, and G. de With, *J. Appl. Phys.*, **75**, 3402 (1994).
12. A. Hiraki, *Jpn. J. Appl. Phys.*, **22**, 549 (1983).

Changes in the Electrochemical Behavior of Silicon after Platinum Deposition and Ionic Bombardment

A. Hervier[a], D. Aureau[a], A. Etcheberry[a]

[a] Institut Lavoisier, Université Versailles Saint Quentin en Yvelines
45 Avenue des Etats Unis, Versailles 78035 France

The electrochemical response of thin Pt films on Si was studied at different stages of mixing between Pt and Si, from fully segregated to fully mixed. This phase change was induced by subjecting the films to helium ion bombardment at energies in the range of 30 keV. The formation of a Pt_xSi phase was confirmed by XPS measurements. As the bombardment dose increases, the voltammogram peaks characteristic of bulk platinum eventually disappear, giving way to the practically featureless voltammogram of electrocatalytically inactive platinum silicide.

Introduction

The electrochemistry of Si has been extensively studied, because it is a ubiquitous material in the electronics industry, and also because of its selective etching properties, which make it possible to create pores with specific geometries.[1] Much less is known, however, about the electrochemical behavior of Si when it is covered by a metal film, despite how common this configuration is in fields such as photovoltaics and micro-electronics.[2,3] We investigated the case of Pt films on Si, and in particular the changes in the cyclic voltammetry of these films when phase mixing is induced between Pt and Si. This was done by exposing the films to a beam of helium ions with energies in the range of 30 keV. By applying different doses of ion bombardment, it was possible to form platinum silicides in a range of stoichiometries, and to observe how this change affected the electrochemistry of the film. Initially similar to bulk platinum, the voltammogram remains mostly unchanged, until the film has almost completely transitioned to PtSi, at which point most of the electrocatalytic activity has disappeared, as is characteristic of platinum silicide.[4]

Experimental Section

The samples used consisted of 10 nm Pt films deposited onto polished p-type Si wafers with a resistivity of 5 mΩ.cm. The bombardment was performed by Quertech, in Caen, France. Helium ions are generated by exposing helium gas to a high frequency oscillating electric field. Magnetic confinement of the ions helps increase the efficiency of the ionization process. The ions are accelerated by an electric field onto the surface of the Pt / Si samples at an incidence angle of 90°, and with an energy of 30 keV.

The doses reported here are proportional to the number of ions impinging on the surface, but are given in arbitrary units. Two batches of samples were treated separately, the first with doses ranging from 250 to 8000, and the second from 250 to 4000.

XPS measurements were performed in a Thermo Scientific K-Alpha spectrometer using a monochromatic Al-Kα X-Ray source (1486.6 eV). Electrochemical measurements were performed in 0.5 M H_2SO_4, using an Ag/AgCl reference electrode and a platinum counter electrode. The samples are put in contact with the electrolyte by pressing them against a Teflon gasket, located on the side of the electrochemical cell, using a metal plate which also serves as a back-contact to the sample. The sample area exposed to the electrolyte is 1 cm^2. Cyclic voltammograms were taken at a scan rate of 50 $mV.s^{-1}$ using a Parstat 2273 potentiostat. During these measurements, the electrochemical cell was placed inside a grounded Faraday cage, and the sample was shielded from any exposure to light.

Results and Discussion

Figure 1 shows the Pt4f XPS spectra for three samples treated with different doses of helium ion bombardment. In order to better visualize the chemical changes undergone by the Pt, all of these spectra have been normalized with respect to intensity. In reality, and as is expected, the ion bombardment leads to a decrease of the intensity of the Pt XPS signal and an increase of the Si signal. It is however difficult to determine whether the metal is ablated from the surface or buried deeper than the probing depth.

For a dose of 250 (in arbitrary units proportional to the number of helium ions impinging on the surface), the signal is comprised mostly of a contribution at 71.2 eV (in the Pt4f$_{7/2}$ portion of the spectrum), corresponding to metallic platinum, and a smaller contribution at a higher energy, 71.75 eV. At an intermediate dose of 2000, the high energy contribution has grown relative to metallic platinum, and has shifted upward slightly in energy, to 72.4 eV. By 8000, the low energy peak has almost disappeared, leaving only the high energy peak.

This behavior has been observed before for Pt/Si films subjected to ion bombardment, and has been attributed to the formation of PtSi, on the basis, for example, of X-ray diffraction and electron diffraction measurements.[5-7] The Pt4f spectra obtained for intermediate doses, where both contributions are still present, are in fact quite similar to those of PtSi films obtained by thermal annealing in reducing conditions.[8] The higher energy peak in the Pt4f spectrum can therefore be assigned to a Pt$_x$Si phase formed under bombardment.

Regardless of whether the silicide film is obtained by annealing or ion bombardment, mixing proceeds through a gradual stoichiometric transition, from a Pt rich silicide, Pt$_3$Si, to PtSi. This transition coincides with an upward shift in binding energy of the silicide peak. The spectra obtained for our samples likewise show that, as the high energy peak grows, it also shifts to higher energy, as shown in Figure 2, confirming the formation of silicide at the surface.

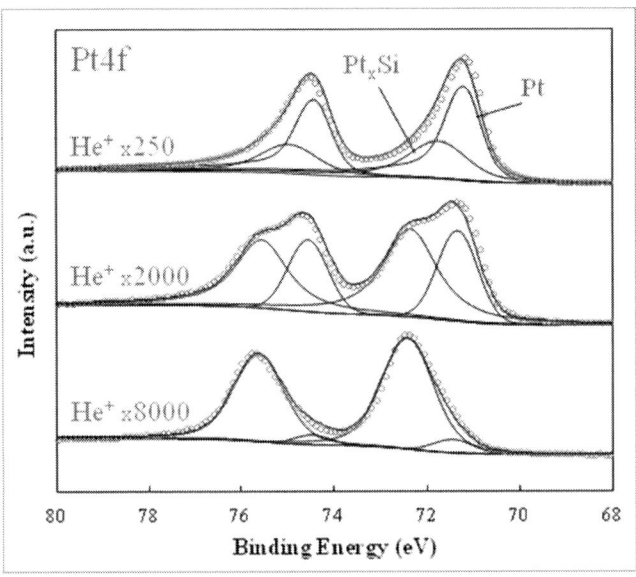

Figure 1. Pt4f XPS spectra of Pt/Si films exposed to increasing doses of helium ion bombardment. The spectra shown here are normalized with respect to intensity.

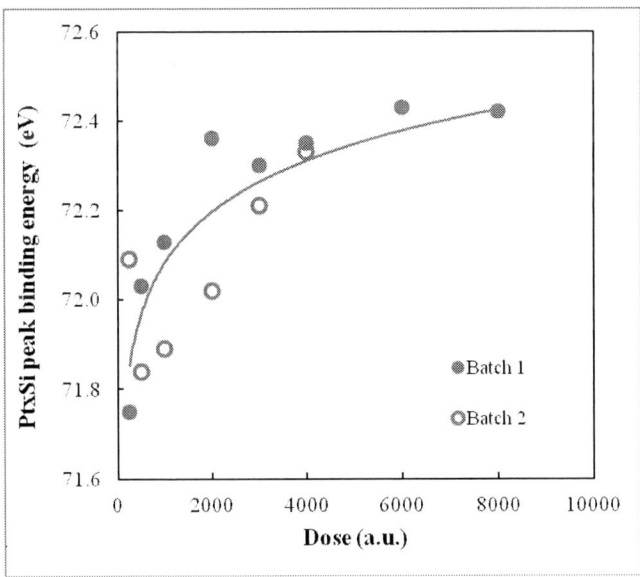

Figure 2. Binding energy of the high energy peak in the Pt4f spectrum (assigned to Pt_xSi) as a function of the helium ion bombardment dose. The line is shown to guide the eye.

Having established the formation of a platinum silicide phase, we can observe the shift in the electrochemical behavior of the film as the mixing occurs. Figure 3 shows the open circuit potential of the samples as a function of the ion bombardment dose. As the helium dose increases, the open circuit potential drops significantly, from roughly 0.7 V for a dose of 250, to 0.4 V at a dose of 8000.

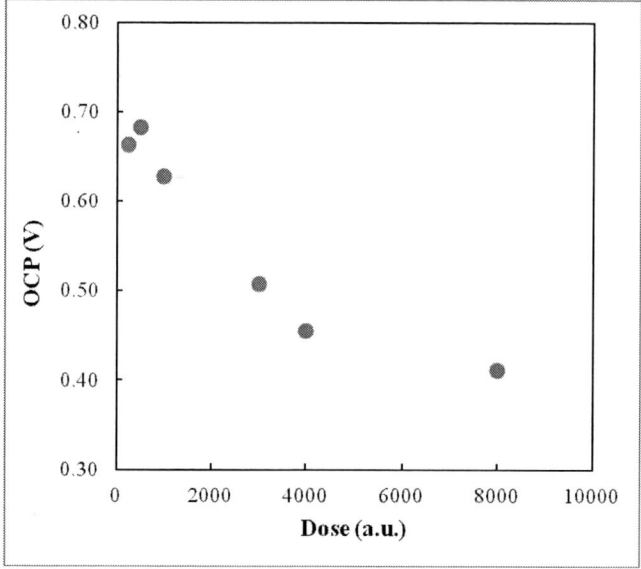

Figure 3.   Open circuit potential of Pt/Si films as a function of the helium ion bombardment dose.

Cyclic voltammograms were also measured. Because the samples were exposed to air for a long period of time between the bombardment and the electrochemical measurements, their surface was heavily contaminated, and initial voltammograms provide very little useful information. One common method of cleaning a surface is flame annealing, which is excluded in our case, since it might wash out the effects of the helium bombardment.[9] Another method consists of simply performing several oxidation-reduction cycles until contaminants have been degraded.[10]

For low dose samples, these cleaning cycles reveal the voltammogram shown in Figure 4, which is similar to what is obtained on bulk platinum.[11] Most characteristically, we observe the peaks attributed to H uptake and desorption, as well as the Pt oxidation and reduction peaks.

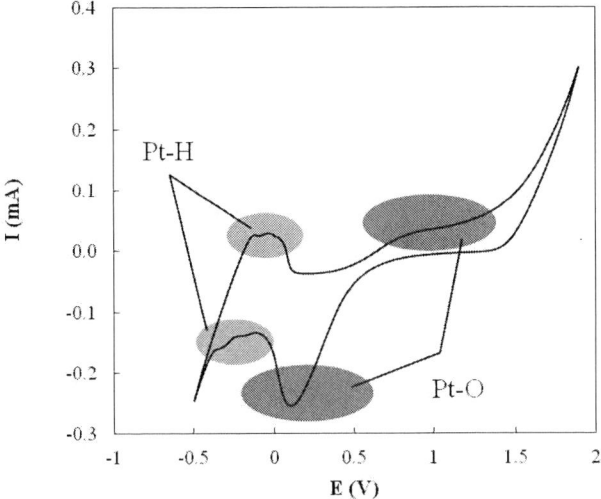

Figure 4. Cyclic voltammogram of a Pt/Si sample subjected to a helium ion bombardment dose of 250. Highlighted in blue and red are the hydrogen uptake/desorption and the platinum oxidation/reduction peaks, respectively.

Recently, Bürstel et al. reported that the voltammetric response of Pt thin films on Si was similar in shape to that of bulk Pt, both when making contact directly to the Pt film (front-contact) and when making contact to the Si substrate (back-contact).[11] The Pt/Si interface, however, is rectifying in the absence of phase mixing, and should therefore present a barrier to electron transport, preventing the measurement of a voltammogram in a back-contact configuration. Our result therefore seems to confirm Bürstel's conclusion, namely that the current measured in these cyclic voltammetry experiments represents a flow of electrons with energies in excess of the Fermi level in Pt.

At doses of 3000 and 4000, the cyclic voltammograms are still overwhelmingly dominated by the response of the platinum (Figures 5b, and c), even though the XPS spectra clearly show that a significant fraction of the surface is comprised of PtSi. This may be in part due to the higher conductivity of the metallic Pt portion of the surface, which conducts more readily. Another possible reason for this phenomenon is that the Pt_xSi phase is likely to form at the buried Pt/Si interface, and therefore have little effect initially on the voltammogram.

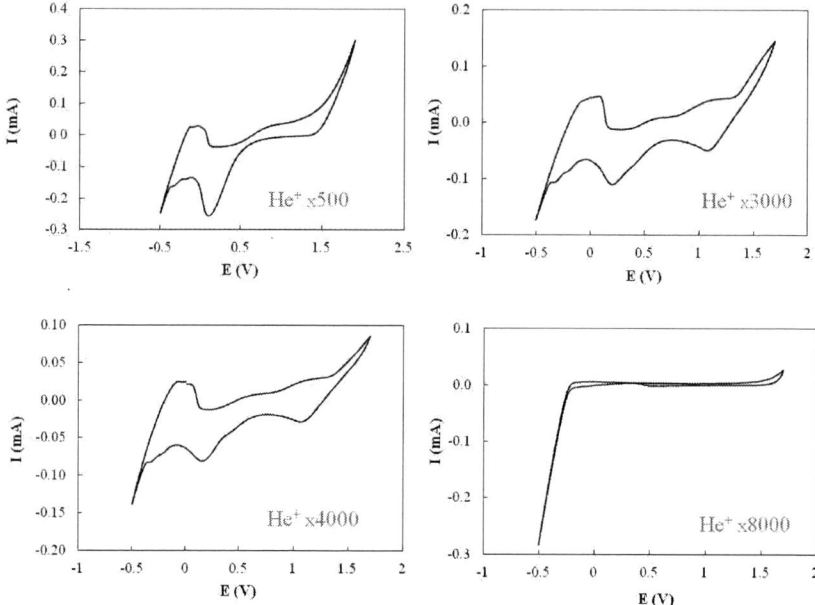

Figure 5. Cyclic voltammograms of Pt/Si films from batch 2 for different doses of helium ion bombardment, measured using an Ag/AgCl reference electrode. The shapes of voltammograms (a), (b), and (c) are characteristic of bulk platinum, while the absence of features in voltammogram (d) is consistent with the lack of electrocatalytic activity on PtSi.

At a dose of 4000, the PtSi in the XPS Pt4f spectrum represents roughly half of the platinum signal, yet the voltammogram shows very little change compared to those at lower doses. However, for a dose of 8000, where the PtSi XPS contribution is now over 90%, we obtain the much flatter voltammogram shown in Figure 4d. The absence of peaks is consistent with the formation of silicide, which has little electrocatalytic activity.[4] If the growth of the Pt_xSi phase is initiated at the Pt/Si interface, changes in the voltammetry may be occurring once the silicide has grown all the way through the platinum film, and is now in direct contact with the electrolyte.

Conclusion

Surfaces of different compositions, ranging from pure Pt films on Si to Pt_xSi, were fabricated by bombarding Pt thin films on Si with helium ions. XPS analysis shows the growth of a feature in the Pt4f spectrum attributable to Pt_xSi. The binding energy of this feature shifts upwards as the fraction of Pt in the silicide increases, consistent with previous reports from the literature. Cyclic voltammetry allows us to observe this transition electrochemically, even when making contact to the Si substrate rather than the Pt film. The voltammograms are similar to those of bulk, polycrystalline platinum, and remain so until the bombardment dose is large enough that the Pt metal contribution has almost completely disappeared from the XPS spectrum, at which point the voltammogram becomes characteristic of Pt_xSi.

Acknowledgments

This work was carried out with support from the *Agence Nationale de la Recherche*. We are grateful to Quertech for performing the ion bombardments.

References

1. V. Lehmann, *Electrochemistry of Silicon: Instrumentation, Science, Materials and Applications, by Volker Lehmann, pp. 286. ISBN 3-527-29321-3. Wiley-VCH, April 2002.*, **1** (2002).

2. C. Wang, J. P. Snyder, and J. Tucker, *Appl. Phys. Lett.*, **74**, 1174–1176 (1999).

3. M. Kimata, M. Denda, N. Yutani, S. Iwade, and N. Tsubouchi, *IEEE J. Solid-St. Circ.*, **22**, 1124–1129 (1987).

4. L. Harris and J. Hugo, *J. Electrochem. Soc.*, **128**, 1203–1211 (1981).

5. B. Y. Tsaur, Z. L. Liau, and J. W. Mayer, *Appl. Phys. Lett.*, **34**, 168–170 (1979).

6. S. Vasiljev, N. Gerasimenko, V. Kalinin, and V. Kulikauskas, *Nucl. Instrum. Meth. B*, **19**, 749–752 (1987).

7. L. Hung, Q. Hong, and J. Mayer, *Nucl. Instrum. Meth. B*, **37**, 414–419 (1989).

8. J. Čechal and T. Šikola, *Surf. Sci.*, **600**, 4717–4722 (2006).

9. J. Clavilier, R. Faure, G. Guinet, and R. Durand, *J. Electroanal.Chem.*, **107**, 205–209 (1979).

10. K. S. Han and O. H. Han, *Electrochim. acta*, **47**, 519–523 (2001).

11. D. Bürstel and D. Diesing, *Phys. Status Solidi (A)*, **209**, 805–812 (2012).

ECS Transactions, 53 (6) 113-121 (2013)
10.1149/05306.0113ecst ©The Electrochemical Society

Preliminary Investigations of Ta Surface Chemistry in Aqueous Solutions of TeO$_2$, and the Possible Formation of TaTe$_2$

Chu Tsang[a], Youn-Geun Kim[a], Daniel Gebregziabiher[a], and John Stickney[a]

[a] Department of Chemistry, University of Georgia, Athens, Georgia 30602, USA

Reductive removal of Ta oxide electrochemically was studied using Cyclic Voltammetry (CV), in-situ electrochemical scanning tunneling microscopy (EC-STM), and X-ray photoelectron spectroscopy (XPS). From CVs, it was shown that the longer a Ta electrode was maintained at negative potentials in excess of -1.5 V, the more surface oxide was reduced, as evidenced by increases in hydrogen evolution. Atomically-resolved EC-STM images were obtained after reduction at -1.8 V, and then imaging at -1.0 V. The efficacy of using Te to passivate the Ta surface was also investigated, and Te appeared to suppress oxide formation, as evidenced by EC-STM, CVs and XPS. Images of the Ta surface in a TeO$_2$ solution suggested the formation of TaTe$_2$ at the surface.

Introduction

TaX$_n$ (X = S, Se, or Te; n = 2 or 3) are transition metal chalcogenides (TMCs) that possess layered structures. Similar to graphene in graphite, the highly-anisotropic TMCs are composed of layers that interact through van der Waals gaps. Interesting phenomena such as charge-density waves (CDW) and superconductivity arise, at least partially, from the compound's anisotropy. (1) Recently there has been heightened interest in application of TMCs in electronic and optoelectronic devices.(2, 3) Attention has focused particularly on 2D nanosheets of the TMCs, which may exhibit properties different from the corresponding bulk materials. Recent experimental and theoretical works have indicated that thin films of transition metal chalcogenides can exhibit CDW transition temperatures higher than those of their bulk counterparts.(4, 5) High-quality TMCs have been traditionally grown by the chemical vapor transport methods,(6-8) which require high temperatures and are time consuming. The need for thin film CDW materials has motivated growing TMCs by alternative methods, possibly by electrochemical atomic layer deposition (E-ALD).

The present study is a first step in determining if E-ALD can be used to grow metal (valve metals: (Ta, Nb and Ti) chalcogenide (S, Se and Te) films at room temperature. Atomic Layer Deposition (ALD) is a deposition method based on the use of surface-limited reactions to form deposits one atomic layer at a time. Electrochemical surface limited reactions are referred to as underpotential deposits (UPD).(9-11) UPD is where the first atomic layer deposits at a potential prior to, or under, that needed to deposit the element on itself. In E-ALD atomic layers of the elements are deposited alternately on each other, in a cycle. The more cycles performed the thicker the deposit. To develop an E-ALD cycle for the deposition of a compound, the electrochemical behavior of the

113

participating elements on each other should first be investigated. This is a study designed to understand the surface chemistry of Te on Ta, as a precursor to designing a cycle for the formation of TaTe$_2$.

Tantalum is a refractory metal that forms a stable surface oxide layer when exposed to ambient conditions. This passivating oxide is resistant to most commonly-known acids and bases, making further reactions with metallic tantalum difficult. In order to fabricate materials of high quality using E-ALD, or to study the interaction between chalcogenide and Ta, the first step is to remove the surface oxide from Ta. In this study attempts were made to remove the oxide from Ta by electrochemical reduction. To follow the state of the surface under electrochemical reduction, Cyclic Voltammetry (CV), Electrochemical Scanning Tunneling Microscopy (EC-STM) and X-ray Photoelectron Spectroscopy (XPS) measurements were performed. In a second step, deposition of Te on the Ta using a TeO$_2$ solution was performed.

Experimental

All potentials were reported vs. a Ag/AgCl (3 M KCl) reference electrode (Bioanalytical Systems, Inc.). The Ta foils (Alfa Aesar, 99.95%) were 0.5 mm thick. Ta/Au/glass substrates were made by in-house sputtering of Ta (Kurt J. Lesker, 99.95%) by e-beam onto commercial Au/glass (EMF Corp) substrates. The Au/glass substrates had a 100 nm Au layer, and a 5 nm Ti adhesion layer existed between the Au and glass. The e-beam deposited Ta layer was ~100 nm thick. All solutions were prepared with 18 MΩ H$_2$O from a Milli-Q purification system. The following reagents were used: NaClO$_4$ (Fisher Scientific, HPLC grade), NaHCO$_3$ (J.T. Baker), KOH (J.T. Baker), NaOH (Fisher Scientific, 98.9%), TeO$_2$ (Alfa Aesar, 99.995%), KClO$_4$ (99.994).

A Nanoscope III (Digital Instruments, Santa Barbara, CA) using a custom built in-situ, EC-STM cell was used. This cell used a Au wire auxiliary electrode and one of the Ag/AgCl reference electrodes described above. XPS analyses were carried out using a Mg K$\alpha_{1,2}$ (STAIB) source at 54.7° take-off angle and a hemispherical analyzer from Leybold Heraeus.

Results and Discussion

Ta Oxide Reduction Study

The redox behavior of a Ta foil electrode was initially studied using cyclic voltammetry (CV). The inset of Fig. 1 shows a CV cycle of the Ta electrode in 1.78 M KOH over the potential range of -1.6 V and -0.2 V. Two points are observed where the currents of the forward and reverse scans cross, at -1.48 V and -0.55 V. It is proposed that the bulk of the charge passed negative of -0.2 V is due to hydrogen evolution reaction (HER), while that positive of -0.2 V results in Ta oxide formation. It is assumed that more oxide present will result in a greater hydrogen overpotential and lower HER below -0.2 V. From the inset of Figure 1, it is evident that there is less HER scanning negative than on the subsequent positive scan. The increase in HER during the positive scan suggests some oxide loss at potentials near -1.6 V. Scans were performed with the

Ta electrode to various positive potential limits. Figure 1 displays the subsequent HER currents near -1.6 V, after reversing the previous scan at the indicated potentials. When the potential was scanned to -420 mV twice, the second scan shows more HER current than the first scan, suggesting there was net oxide removal. As the positive potential limit was increased, less HER current would result. Evidently, any oxide produced on the positive scan resulted in lower HER currents on the negative scan. If the potential was not allowed over -0.42 V, no significant oxide was formed, and during the second cycle, more of the oxide was reduced at -1.6 V, as evidenced by the increase in HER.

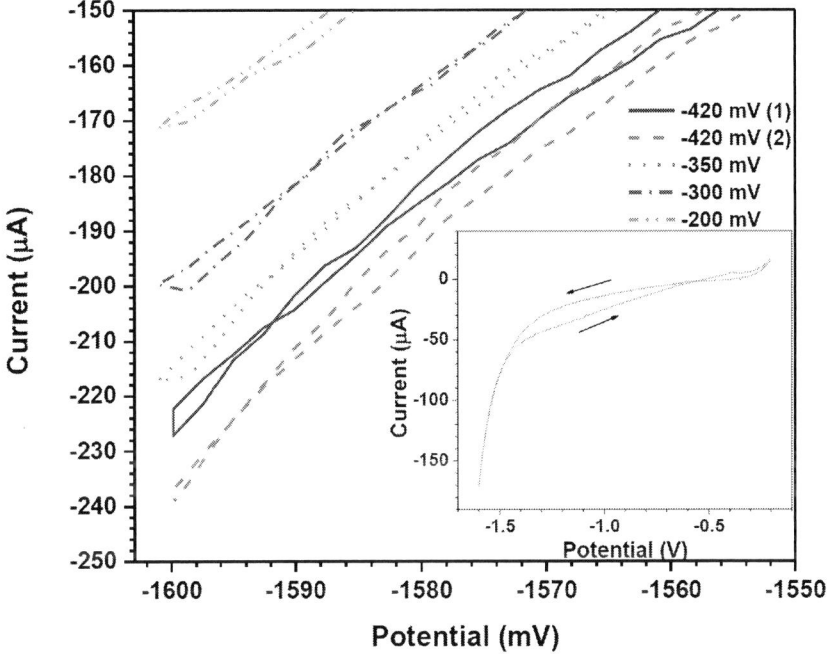

Figure 1. Ta foil in 1.78 M KOH (pH 14). Scan rate: 10 mV/s. Electrode area: 1.8 cm^2.

Figure 2 shows EC-STM images of the Ta electrode surface while it was scanned between -1.3 V and -0.6 V in a 0.1 M KClO$_4$ electrolyte solution. The surface was first held at -1.0 V for 10 min, to reduce Ta oxide, prior to imaging. Patches showing a step-terraced surface, with 5 – 10 nm steps, were observed. Triangular pits, as outlined in the right image of Fig. 2, were also observed and grew during the scan. Atomic-level images of the surface were not obtained, presumably because the surface was not completely reduced. The potential was then held at -1.8 V for approximately 30 min. Due to extensive hydrogen evolution at -1.8 V, the potential was stepped to -1.0 V for imaging, and atomically-resolved images were observed as shown in Fig. 3. The surface appeared to predominantly display the Ta(110) surface, with a (1x1) structure. After approximately 20 min, atomic resolution was lost, presumably because the surface was slowly re-oxidizing at -1.0 V.

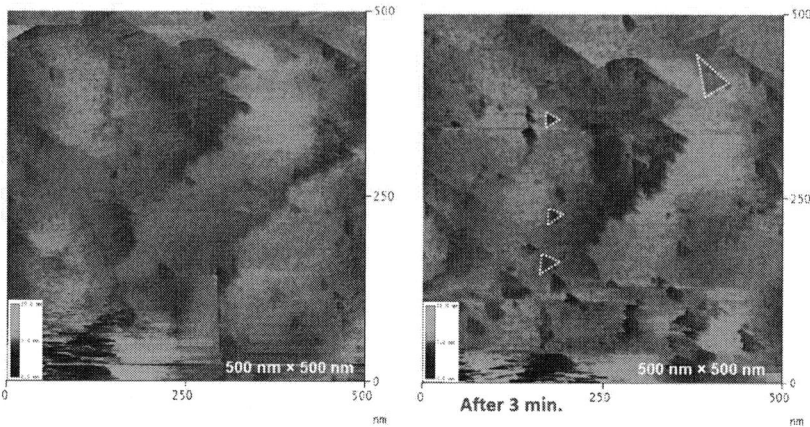

Figure 2. EC-STM images of Ta foil at between -1.3 V and -0.6 V after holding at -1.0 V for oxide reduction. 0.1 M KClO$_4$ (pH ~5) was used as the electrolyte. Triangular pits that grew during imaging are outlined in dashes on the right image.

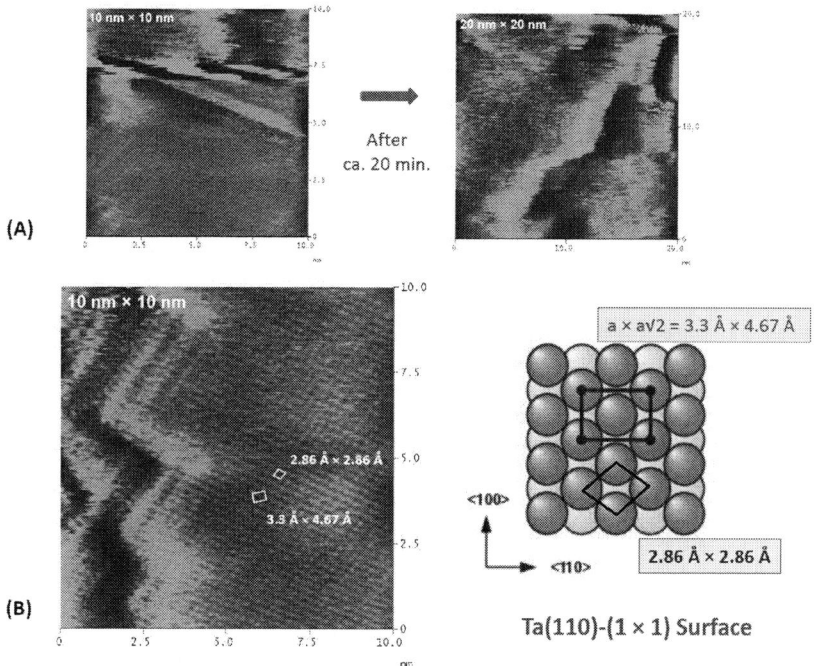

Figure 3. (A) EC-STM images of Ta surface at -1.0 V after the -1.8 V oxide removal step. Atomic resolution was lost after approximately 20 min due to re-oxidation of the surface. (B) A comparison showing the surface is predominantly Ta(110)-(1x1).

Characterization of Te Deposition onto Ta by EC-STM

Te was investigated as a possible passivating agent for Ta, given its ability to passivate other metal surfaces such as Cu and Au, and because the electrochemistry of Te has been study by this group for 25 years.(12, 13) Successful deposition of Te onto Ta is also relevant to the attempts by this group to form metal chalcogenides for CDW device fabrication. After reductive removal of the Ta oxide layer, Te was deposited onto the Ta surface at -1.0 V. Figure 4 shows atomically-resolved EC-STM images of the resulting surface. After Te deposition the Ta surface was relatively easy to image. The images were stable for much longer and over a wider range of potentials, up to at least -0.3 V, compared to the Te-free surface. Both observations suggest the electrodeposited Te layer protected the Ta surface from oxidation.

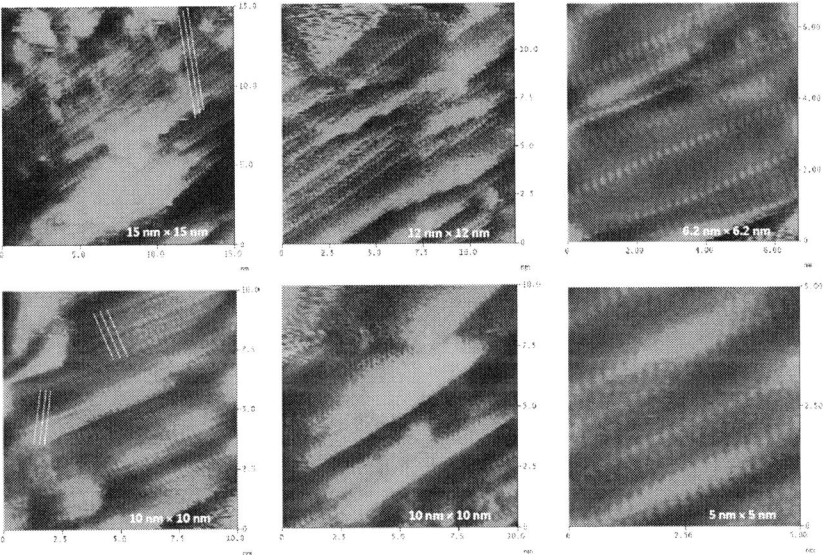

Figure 4. EC-STM images of various densely-packed bcc(hkl) surfaces of Te on Ta/Au/glass. Te was deposited onto a Ta/Au/glass substrate from a solution containing 0.2 mM TeO_2 and 0.1 M $KClO_4$.

A particularly interesting Te-covered surface structure is shown in Fig. 5. The structure resembles a "double zigzag" pattern reported by Kim et al for a 1T-TaTe$_2$ (001) surface, suggesting TaTe$_2$ may have formed on the electrode surface.(14) Kim used AFM to image the single crystal TaTe$_2$ surface and measured the unit cell parameters a and b to be 19.2 ± 0.2 Å and 3.6 ± 0.2 Å, respectively, which are in agreement with the crystallographic data published by Brown. (15) In this EC-STM study, the measured interatomic distance is 3.3 Å and the periodic distance in the direction perpendicular to the rows of Te atoms is about 22 Å as indicated in Fig. 5. The measured distances are almost in agreement with the values reported respectively by Kim and Brown.

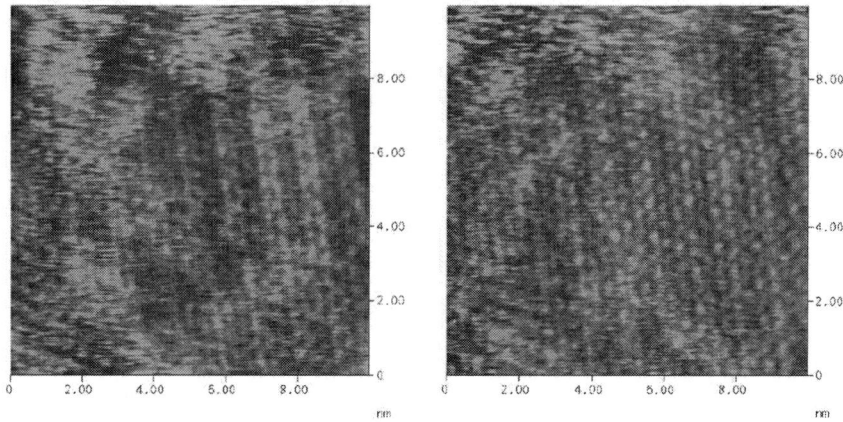

Figure 5. EC-STM images of a Te-covered Ta/Au/glass surface. The atomic distance is 0.33 ± 0.02 Å. The distance indicated by the solid line in the left image is ~22 Å.

If $TaTe_2$ has been formed on the Ta electrode, the discrepancy between the measured distances might be the result of drift in the microscope. If the Te is present as a single layer on the surface of Ta, the resulting position of Te atoms would not be expected to be those of a $TaTe_2$ nanofilm. During imaging, some small particles appear in the images, suggesting that some nm-sized flakes of $TaTe_2$ are present. The particles are very mobile suggesting they are not strongly bound to the surface, and the tunneling process can move them, as would be expected for a van der Waals material like $TaTe_2$.

Characterization of Te Deposition on Ta by CV and XPS

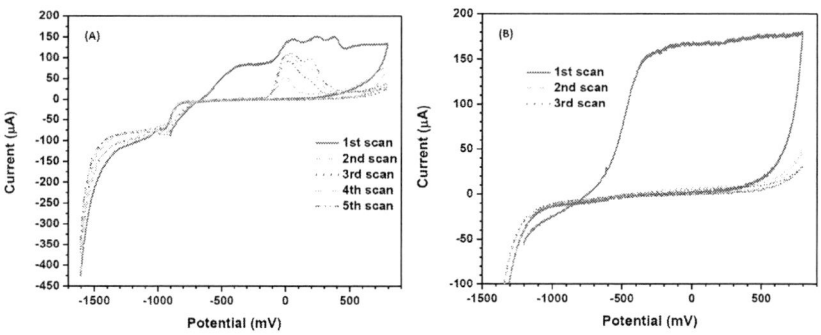

Figure 6. (A) CV of polished Ta foil in 0.25 mM TeO_2 (pH 10.5). The potential was initially held at -0.9 V for 10 min before started scanning positive. (B) CV of polished Ta foil in 0.1 M $NaClO_4$ blank (pH 10.5). The potential was initially held at -1.2 V for 10 min before started scanning positive.

Cyclic voltammetry and X-ray photoelectron spectroscopy (XPS) were used to further verify the deposition of Te on Ta electrodes and the possibility of using Te to passivate a Ta surface. CVs of a Ta foil electrode in a pH 10.5 solution of TeO_2 are shown in Fig. 6A. The potential was initially held at -0.9 V to promote oxide reduction and Te deposition. In the initial positive scan, oxidation was observed above -0.66 V. From the blank scan (Fig. 6b), this large oxidation current was only observed for the first positive scan. It was considered to be the formation of an irreversible tantalum oxide, or oxidation of an unstable layer of the initial Ta surface. (16) On top of this irreversible oxidation current, bulk Te oxidation is observed starting at -0.15 V in Figure 6A. It is interesting that the oxidation current negative of -0.1 V in Figure 6A was less than that above 0.3 V, when compared with Figure 6B. This increased in tantalum oxidation current suggests the Te may have suppressed some of the Ta oxidation. Between -0.1 and 0.3 V in Figure 6A three peaks for Te oxidation are evident. The second cycle, however, shows only one. Subsequent cycles show the other two Te oxidation peaks becoming more prominent. A reduction feature between -1.0 V and -0.7 V corresponds to the initial deposition of Te. The lack of Te features between -0.2 V and -0.8 V is characteristic of the irreversibility of the deposition of Te.

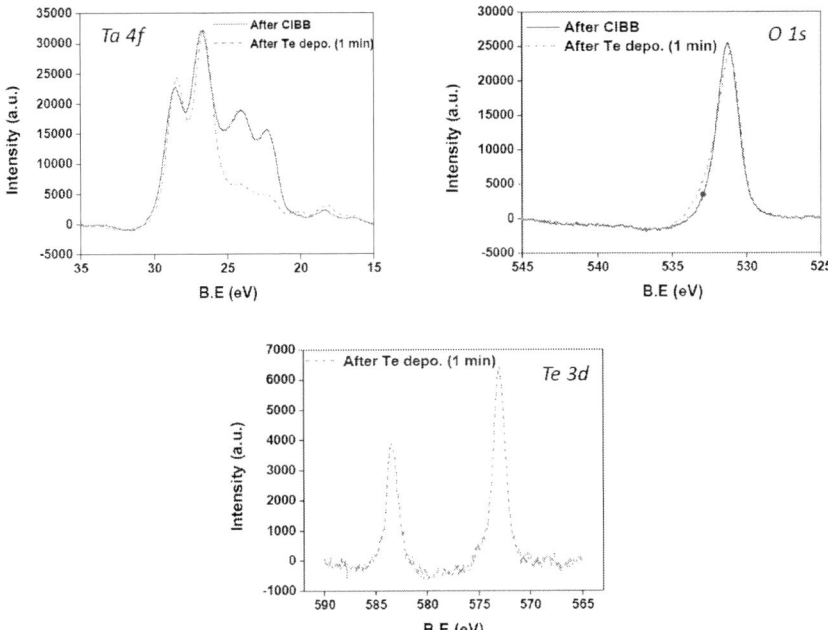

Figure 7. XPS spectra of Ta/Au/glass electrode before and after Te deposition. Te was deposited in a solution containing 0.25 mM TeO_2, 18 mM NaOH and 20 mM $NaHCO_3$. Prior to Te deposition, the electrode was cleaned by cold ion bombardment (CIBB).

A Ta/Au/glass electrode was first cleaned by cold Ar^+ ion bombardment in the UHV chamber, and then transferred to an UHV ante-chamber equipped with an electrochemical cell. The open-circuit potential (OCP) in a pH 10 TeO_2 solution was -0.85 V. Next, Te was deposited by scanning the potential negative of OCP and cycled between -1.6 V and -0.4 V three times, followed by a 1-min hold at -1.6 V. Following the Te deposition, the sample was transferred back to the UHV analysis chamber for XPS analysis. Figure 7 shows the XPS data and evidence of the presence of Te after this deposition. Spectra of Ta and O before and after Te deposition are also shown for comparison. While the intensities of the oxidized Ta peaks remain relatively unchanged, the metallic Ta peak intensities drop significantly, indicating probable oxidation of the metallic Ta near the surface. However, the O signal intensity is relatively unchanged, possibly indicating that the metallic tantalum reacted instead with Te, possibly protecting the Ta from oxidation.

Conclusion

The electrochemical oxide reduction of Ta was studied by CV, STM and XPS. CV initially suggested some oxide can be reduced at very negative potentials, as indicated by changes in the HER current. In the EC-STM study, an oxide-removal step at -1.8 V was used to obtain atomic resolution images. The EC-STM study also showed that an oxide-free Ta surface would spontaneously re-oxidize even at -1 V. Use of Te to passivate the Ta surface also resulted in atomic resolution, and suggested greatly increased stability, up to as much as -0.3 V. A "double zigzag" pattern observed in some STM images appears to be evidence of $TaTe_2$ formed on the surface. CV and XPS data support these conclusions, but there are more questions than answers in this report. This work is continuing

Acknowledgments

Support for this work is gratefully acknowledged and was from NSF ECCS, award number 1124733.

References

1. J. A. Wilson, F. J. Disalvo and S. Mahajan, *Phys Rev Lett*, **32**, 882 (1974).
2. N. Ogawa and K. Miyano, *Appl Phys Lett*, **80**, 3225 (2002).
3. D. Mihailovic, D. Dvorsek, V. V. Kabanov, J. Demsar, L. Forro and H. Berger, *Appl Phys Lett*, **80**, 871 (2002).
4. P. Goli, J. Khan, D. Wickramaratne, R. K. Lake and A. A. Balandin, *Nano Letters*, **12**, 5941 (2012).
5. M. Calandra, I. I. Mazin and F. Mauri, *Phys Rev B*, **80** (2009).
6. F. Levy and H. Berger, *J Cryst Growth*, **61**, 61 (1983).
7. R. E. Thorne, *Phys Rev B*, **45**, 5804 (1992).
8. A. J. Patel, M. K. Bhayani and A. R. Jani, *Chalcogenide Lett*, **6**, 491 (2009).
9. R. Vaidyanathan, S. M. Cox, U. Happek, D. Banga, M. K. Mathe and J. L. Stickney, *Langmuir*, **22**, 10590 (2006).
10. V. Venkatasamy, N. Jayaraju, S. M. Cox, C. Thambidurai, M. Mathe and J. L. Stickney, *J Electroanal Chem*, **589**, 195 (2006).
11. T. E. Lister and J. L. Stickney, *Appl Surf Sci*, **107**, 153 (1996).

12. D. K. Gebregziabiher, M. A. Ledina, R. Preisser and J. L. Stickney, *J Electrochem Soc*, **159**, H675 (2012).
13. M. D. Lay and J. L. Stickney, *J Electrochem Soc*, **151**, C431 (2004).
14. S. J. Kim, S. J. Park, I. C. Jeon, C. Kim, C. Pyun and K. A. Yee, *J Phys Chem Solids*, **58**, 659 (1997).
15. B. E. Brown, *Acta Crystallogr*, **20**, 264 (1966).
16. S. B. Emery, J. L. Hubbley and D. Roy, *J Electroanal Chem*, **568**, 121 (2004).

The Nanoporous Metallisation of Polymer Membranes through Photocatalytically Initiated Electroless Deposition

M. A. Bromley and C. Boxall

Engineering Department, Lancaster University, Lancaster, LA1 4YR, UK

We present the novel use of Photocatalytically Initiated Electroless Deposition (PIED) for the deposition of metal films with highly ordered arrays of sub-μm (hemi)spherical pores directly onto the surface of insulating organic membrane-based substrates. This is achieved by sensitisation of the target substrate with a TiO_2 photocatalyst followed by the self-assembly of a hexagonally close packed polystyrene microsphere template at the substrate surface. Metallisation then occurs through PIED into the template interstices and directly onto the TiO_2 sensitised membrane surface. The dimensions of the resultant pores in the deposited metal are determined by the size of the template microspheres while metal film thickness may be controlled by the deposition period.

The fabrication of nanoporous metal by this novel method adds a conductive and permeable metallic structure of high surface area to an otherwise electrically insulating polymer membrane surface. Such metallised insulating membranes have potentially wide applications in membrane and separation technology, desalination and electrode / solid electrolyte composites for fuel cells.

Introduction

Fast, controllable and selective separation of metal ions from complex solutions is key to a wide range of activities including pre-analytical separation, environmental monitoring, environmental remediation, metal recycling, desalination and process control. These activities are important in a wide range of industries including pharmaceuticals, mining, foodstuffs, wastewater processing, power generation and nuclear. Using our recently reported insulator metallisation process of Photocatalytically Initiated Electroless Deposition (PIED) (1-4), we aim to develop novel, highly ion selective membrane-based separation technologies through a combination of supported ligand membranes and supplemental electrochemical control by overlying nano-engineered metal layers. Layers of porous metal with controllable pore size, distribution and metal thickness deposited on ion selective membranes (ISMs) may act as *in-situ* electrodes offering electrochemical control over the environment at the ISM-solution interface. The result is an ability to manipulate the valence state of target ions at the ISM surface, providing greater selectivity in extraction than achieved with a supported ligand membrane-only system.

The principles of PIED have been described in detail elsewhere (1-4) but may be summarised as follows. Metal deposition is initiated through a semiconductor photocatalysis based process whereby, initially, electron-hole pairs are generated in TiO_2 particles upon the absorption of ultra-band gap light energy (λ = 312 nm). These photogenerated electrons may then reduce metal precursor ions in solution (5), forming metal nuclei on the semiconductor surface, Figure 1 (i). Once sufficient nucleation sites have been photocatalytically generated, these begin to grow through an auto-catalytic electroless deposition reaction, driven by a reductant present in the electroless plating bath, Figure 1 (ii); the growing nuclei eventually coalesce to form a coherent metal layer. This second step can occur both in the presence and absence of light, the latter being particularly advantageous for producing thick, non-transparent metal layers.

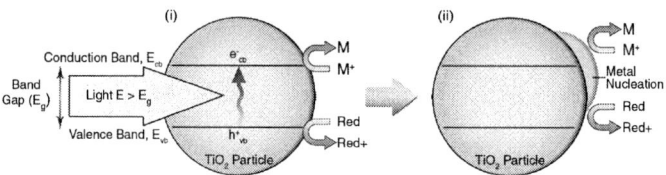

Figure 1 Two stages of PIED: (i) photoreductive metal nucleation on irradiated photocatalyst; (ii) auto-catalytic electroless deposition process

In this basic form, the resulting metal film has no artificially induced structure. In order to produce a metal film with a nanostructure / nanoporosity, the metal deposition must take place around some form of template. We have deposited metal via PIED into the interstitial spaces of hexagonally close packed polystyrene (PS) microsphere arrays in a similar manner to that reported by Bartlett *et al* (6) wherein porous metal films are prepared by electrodeposition into the interstitial spaces between PS spheres assembled on metal electrode surfaces. In each case, the microsphere template is removed post-deposition, by dissolving in toluene, to leave a metal film with a regular nanostructure. Furthermore, polystyrene spheres are commercially available in sizes ranging from 50 nm to 3 μm, allowing for the production of metal films with a wide range of pore sizes using the same deposition technique as demonstrated in parallel work from this laboratory (7).

Having previously developed PIED for the deposition of non-porous and porous metals onto glass (1, 4), the same fabrication techniques can be utilised for the metallisation of other insulating substrate materials. As the TiO_2 sensitisation of the substrate material is of prime significance to the deposition process, theoretically, any sensitised substrate should be receptive to subsequent photocatalytically initiated electroless deposition. However, this sensitisation process on glass usually involves annealing of the TiO_2 coating at 773 K to produce a robust layer of the anatase crystal phase (1), known to display the highest photocatalytic activity (8). As the anatase phase is formed at temperatures from 623-823 K, it is likely that any such thermal treatment will cause damage to many potential organic polymer membrane-based substrate materials.

Hence, this paper concerns the development of PIED for membrane metallisation using a porous polyvinylidene fluoride (PVDF) substrate as an exemplar ISM material.

Experimental

Materials and Reagents

All reagents used are AnalaR grade or higher, and purchased from Sigma Aldrich (Gillingham, Dorset, UK) or Alfa Aesar (Heysham, Lancashire, UK) with the exception of nanoparticulate TiO_2 sol (TiPE® O502, TitanPE Technologies Inc., Shanghai, China). All water used is Ultrapure doubly de-ionised water from a Direct-Q 3 UV Millipore water purification system (Millipore (U.K.) Limited, Watford, UK) to a resistivity of 18.2 MΩ.cm. Nitrogen Whitespot grade is provided by BOC Ltd., Guildford, Surrey, UK.

TiO$_2$ Sensitisation of Substrates Using Nanoparticulate Colloids

PVDF membranes, 0.22 µm pore size (Millipore), were sensitised by coating with a nanoparticulate TiO$_2$ sol (TiPE® O502, TitanPE Technologies Inc., Shanghai, China) deposited via spin-coating for 5 seconds at 2900 rpm using an inverted model 636 rotating disk electrode system (Princeton Applied Research, Tennessee, USA) for single side sensitisation and by manual dip-coating for two sided sensitisation. The sol contains nanoparticulate anatase TiO$_2$ with average primary particle size of < 8 nm in water based solution. Coated PVDF membranes were oven dried at 373 K to remove solvent and so produce a coating of anatase nanoparticles. Sensitised substrates were stored in darkness at room temperature prior to use.

Polystyrene Microsphere Array Formation

Polystyrene Microspheres of 1µm diameter were purchased from Alfa Aesar as a 2.5 wt% solution in water. For template preparation purposes this was further diluted to 1.0 wt% with doubly deionised water. TiO$_2$ coated surfaces exhibit super-hydrophilicity upon irradiation with UV light (9), a phenomenon that we have explored for optimising the self-assembly of ordered microsphere arrays. In this, TiO$_2$ sensitised substrates were irradiated with UV light for 60 min prior to the application of the microsphere solution, the photo-induced substrate hydrophilicity facilitating an even spreading of that solution (see below). 12.35 µl of the microsphere solution was applied to the substrate via micro-pipette inside an 11 mm diameter PTFE ring with tapered inner edge, Figure 2. This ring was sealed to the substrate surface with silicone grease to ensure there is no solution leakage while the tapered edge creates a permanent concave meniscus on the liquid surface, essential for the ordered template self-assembly process, again shown in Figure 2. Upon application of the sphere solution, substrates were left undisturbed for 12 hours on a level surface (to prevent gravitational bias) and in an enclosed environment (to ensure no uneven evaporation is caused by local air currents). Ordered template self-assembly then occurs through the capillary action of the surface tension force of the evenly evaporating liquid meniscus on the component microspheres.

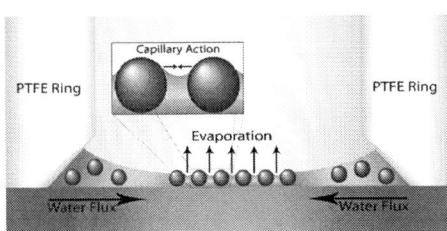

Figure 2 Self-assembly of polystyrene microsphere template

Preparation of Ag Electroless Plating Baths

Electroless Ag plating solutions were prepared to the composition given in Table 1. All components are added to a small quantity of distilled water in the order listed, ensuring full dissolution with each addition. The completed solution was made up to volume with doubly deionised water and purged with N$_2$ for 20 min to deoxygenate. The pH of the solution was 11.5 and PIED was carried out at 298 K. Electroless plating solutions are freshly made immediately before use for optimum performance. As silver nitrate is light sensitive Ag solutions are prepared and stored in amber flasks in darkness.

Role	Component	Concentration
Metal precursor	Silver Nitrate	1.496 g / dm^3 (8.8 mmol / dm^3)
Complexant	Ethylenediamine	3.245 g / dm^3 (54 mmol / dm^3)
Stabiliser	3,5-diiodotyrosine	0.017 g / dm^3 (39.2 μmol / dm^3)
Reducing agent / scavenger	Potassium Sodium Tartrate	0.7356 g / dm^3 (26 mmol / dm^3)
pH	-	11-12
Temp	-	298 K

Table 1 Composition of Ag electroless plating solution

Photocatalytically Initiated Electroless Deposition

TiO$_2$ sensitised, microsphere template coated substrates were placed directly into freshly prepared electroless plating solutions in a quartz reaction vessel for optimal UV transmittance. Plating solutions were bubbled with N$_2$ during deposition in order to purge the solution of oxygen which can compete with the metal precursor ions for reduction at the photocatalytic surface. The N$_2$ stream also provides a source of agitation to prevent local depletion of the metal ion concentration at the substrate surface. The reaction vessel is then placed inside a photoreactor (Lidam Scientific, Dartford, UK) comprised of two hemi-cylinders, each containing 6 x 8W UVA lamps. A thermostatically controlled water supply, passed through the incorporated jacket of the quartz vessel by means of a peristaltic pump, allowed the plating solution to be maintained at a constant temperature. The immersed substrates were exposed to UV light for varying irradiation periods during which PIED occurs.

Polystyrene Microsphere Array Removal

Post-PIED, removal of the microsphere template was achieved by immersing the substrate in a toluene bath to dissolve the polystyrene spheres. Dissolution is rapid, with typical immersion times being in the order of 30 seconds leaving a structure of ordered (hemi) spherical voids within the deposited metal; no adverse effects are observed on the deposited metal or PVDF membrane substrate.

Metal Deposit Characterisation

Morphology and dimensions of the deposited metal layers were determined by scanning electron microscopy (FEI Phenom, Lambda Photometrics Ltd, UK).

Results and Discussion

With a melting point of ~443 K, PVDF is not able to withstand the high annealing temperatures (773 K) applied during the TiO$_2$ sensitisation of more robust substrates such as glass using sol-gel derived titanias (1, 4). Due to this necessity to lower the firing temperature, these sol-gel derived coatings would have to be used in a photocatalytically less active, non-crystalline state (10). Thus, an alternate means of providing a TiO$_2$ coating was explored by using a commercially available / proprietary nanoparticulate

TiO$_2$ suspension sourced from TitanPE Technologies Inc. in Shanghai, China. This suspension contains primarily anatase particles in its stock form. The use of TiPE TiO$_2$ allows the sensitisation of substrates with predomonantly anatase TiO$_2$ without the need for high temperature annealing.

The surface of an unmodified PVDF membrane is shown in Figure 3 (i), material porosity and irregularity being clearly visible. When sensitised by spin-coating, the majority of the applied TiO$_2$ suspension is rapidly removed by centrifugal force leaving only a thin film which closely follows the surface structure of the PVDF surface, Figure 3 (ii). While the deposited TiO$_2$ occludes some smaller features of the porous membrane, the structure remains largely recognisable and the porosity remains unobstructed on a larger scale. Sensitisation by dip-coating increases the TiO$_2$ loading, so enhancing the photocatalytic effect, but the photocatalyst is much more obstructive to the porosity of the PVDF membrane as a consequence, Figure 3 (iii). This occurs because, unlike with spin-coating, TiO$_2$ particles are not discarded with the excess solvent by centrifugal force; oven drying at 373 K removes only the volatile solvent while the TiO$_2$ particles remain on the substrate as an evaporation residue. Only part of the membrane porosity is lost, a factor of significance for the ion separation membranes under consideration here, and, importantly for PIED the concentration of TiO$_2$ at the membrane surface is much greater than that at a similar substrate sensitised by spin-coating.

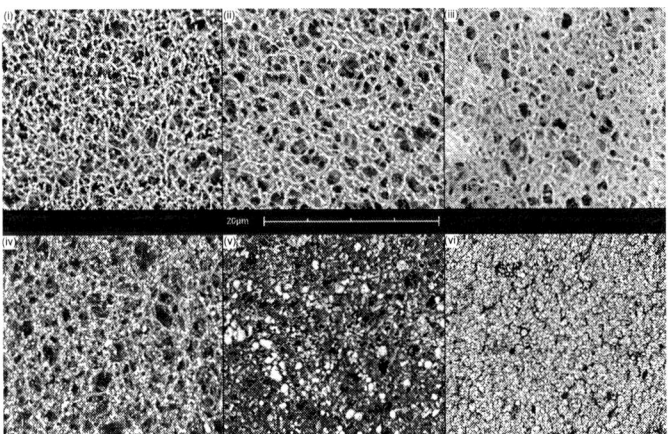

Figure 3 SEM images of (i) un-modified PVDF membrane, (ii) TiO$_2$ loading on PVDF after spin-coating, (iii) TiO$_2$ loading on PVDF after dip-coating; TiO$_2$ sensitised PVDF membrane with (iv) Ag nucleation, t = 15 min, (v) continued Ag growth, t = 30 min and (vi) non-porous, conducting Ag film, t = 60 min

Once suitably sensitised with TiO$_2$, the PVDF membrane can be metallised through PIED. Metal deposition occurs via a number of distinct growth stages, as previously observed with PIED on glass substrates (3). Photocatalytic reduction of metal ions from the electroless plating solution results in metal nucleation directly onto the TiO$_2$ particles. Figure 3 (iv) shows Ag nucleation, visible as bright white particles across all areas, on TiO$_2$ sensitised PVDF after a deposition time of 15 min; at this stage the Ag deposit takes shape around the structure of the PVDF membrane, the porosity of which remains largely un-obscured with the formation of a coherent metal film yet to occur. As the deposition period continues, auto-catalytic electroless deposition, driven by the reductant in the electroless plating bath, begins to occur at each photogenerated nucleation site resulting in the growth of larger Ag granules. For the system of Figure 3, this occurs after approximately 15 – 30 min, Figure 3 (v).

After a total deposition period of 60 min, these isolated metal granules are seen to coalesce and form a coherent and electrically conductive metal film, Figure 3 (vi). SEM analysis reveals that this metallic Ag layer is highly granular, more so than similar layers formed on glass substrates, attributed to the irregularity of the substrate surface and potentially a reduced efficiency of the non-annealed photocatalyst. While a slower electroless deposition is typically expected to produce a deposit of finer granularity, here we see a photocatalytic initiation stage producing fewer nucleation sites than previously observed on m-TiO$_2$ sensitised glass substrates within the same deposition time (3, 4, 7). This, combined with a translation of substrate morphology into the deposited metal results in a rougher and more granular film. Despite this, the metal film displays good electrical conductivity with a sheet resistivity, R_s, of 0.5 Ω/sq, a value comparable to the R_s of 0.2 Ω/sq observed on fully conductive Ag layers on glass (4, 7).

As the dip-coating sensitisation method results in TiO$_2$ deposition onto all membrane surfaces, including the membrane edges, metal deposition also occurs in the same areas resulting in conductivity between opposite sides of the metallised membrane. However, by simply trimming the metallised edges from the membrane, the electrical connection between these opposing membrane sides is eliminated. This trimming procedure then results in the production of two individually conductive metal layers / electrodes on a single porous PVDF membrane with zero through membrane conductivity. Each deposited metal electrode could therefore be separately controlled, thus offering independent electrochemical control of the solution environment on either side of the membrane. Such a phenomenon could be exploited in enhanced separation processes on ISMs if the deposited electrodes themselves were to be rendered porous. This can be achieved via PIED through the interstices of a hexagonally close packed microsphere array that is subsequently removed. This method forms the subject for the remainder of this paper.

Microsphere Array Formation

As previously described in our studies of the metallisation of glass substrates with nanoporous metal films (4, 7), hexagonally close packed arrays of 1 μm microspheres provide the template material around which PIED generated nanoporous metal film may be formed. We have found that such microsphere array templates may also be deposited onto the PVDF membrane substrates in the same manner as on glass, although the spheres exhibit less regular packing on membranes than on similarly prepared glass substrates (4, 11). This may be attributable to two factors; firstly, the rough surface morphology of the PVDF membrane, Figure 3, disrupting the self-assembly process of the microspheres, leading to an increased number of dislocations within the array; and secondly, the potentially reduced efficiency of the non-annealed photocatalyst resulting in the super-hydrophilicity of the irradiated TiO$_2$ surface (9, 12) being less readily generated and the microsphere suspension spreading less freely over the substrate surface. While the latter may be true for TiO$_2$ sensitised membranes in comparison to TiO$_2$ sensitised – and fully annealed – glass substrates, the photocatalytically induced hydrophilic effect achieved on an irradiated, TiO$_2$ sensitised membrane remains advantageous over a non-irradiated, TiO$_2$ sensitised membrane over which the microsphere suspension spreads much less readily due to the lack of hydrophilic assistance.

Due to the aforementioned factors, microsphere templates deposited onto PVDF membranes lack substantial areas of hexagonal close packing as the self-assembly mechanism required for the formation of highly ordered close packed microsphere arrays is disrupted, Figure 4. Despite this, the microspheres are generally deposited within close proximity and the templates are still appropriate for the induction of widespread nano-scale porosity in subsequently deposited metals – though the porosity of the resultant metals will be expected to exhibit a lower degree of regularity on membranes than on glass substrates. This is discussed in more detail below.

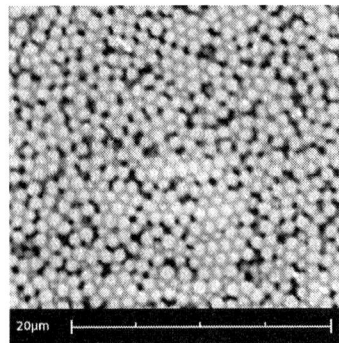

Figure 4 1 μm polystyrene microsphere array on TiPE O502 sensitised PVDF membrane

Templated Ag PIED on PVDF

Microsphere template coated PVDF substrates are metallised through PIED using the same method as described above. After a short 30 min period of irradiated immersion in the Ag electroless plating bath, isolated Ag deposits are observed, Figure 5, and can be seen to form via the same initial growth stages as observed for both templated Ag and Pd PIED on glass substrates (4, 7). Initial nucleation occurs around the base of the microspheres with larger islands of Ag forming through lateral growth of the metal deposit. Interestingly, the pores formed in the metal deposit at the point of contact between the microspheres and the substrate can display an irregular size and shape, some being elongated and not wholly comparable to the highly circular pores observed on glass (4, 7). This is attributable to the morphology of the PVDF membrane, resulting in less uniform areas of contact between individual microspheres and the substrate. It is noticeable that fewer photocatalytically generated nucleation sites occur on membranes with a microsphere template in place, Figure 5, than without, Figure 3 (v) within a comparable time period; this is due to the microspheres reducing the intensity of UV light reaching the photocatalyst. Despite this, enough nucleation sites are still generated to allow auto-catalytic electroless deposition to occur and Ag islands to develop.

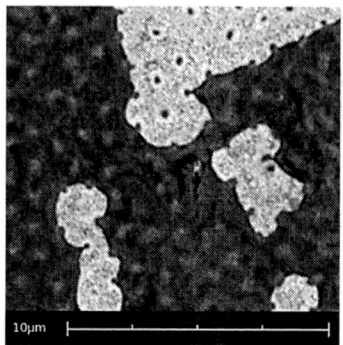

Figure 5 Ag nucleation and island growth around 1 μm microsphere template on PVDF, t = 30 min

Lengthening the deposition period to 60 min enables further metallisation and the formation of a conductive Ag deposit, with a sheet resistivity of 0.5 Ω/sq over the entire templated area, Figure 6. All deposited Ag remains adhered to the substrate upon removal of the microsphere template by toluene dissolution, with no delamination occurring during dissolution or drying stages. The underlying membrane also appears impervious to any detrimental effects from the microsphere removal process, with no shrinkage, melting or other deformation evident. Assessment of the adhesion strength between the deposited metal and substrate poses difficulties on flexible polymer membranes such as PVDF; the British Standard metal adhesion test BS EN ISO 2819:1995, the so-called "tape test", is unsuitable as the method causes tearing of the membrane without properly examining the adhesion of the metal layer. However, the fact that both the sensitising TiO_2 film and the deposited metal are able to penetrate the porous structure of the PVDF membrane and provide a complex, inter-locking interface between the deposits and the substrate suggests that a strong adhesion should be formed.

The SEM analysis of Figure 6 confirms that nanoporosity is successfully induced into Ag on a PVDF membrane support. The film is seen to be comprised of a monolayer of hemispherical voids in the metal matrix. While less ordered than previously observed in analogous samples on glass (4, 7), regular porosity is present across the sample. The diameter of the mouths to the hemispherical pores / voids is found to be 690 nm. From this, a metal layer depth of 138 nm can be calculated through geometric determination of the sagitta of the hemispherical void; as these measurements are found to be consistent across the sample area, it can be concluded that metal deposition has occurred in a homogenous manner with uniform thickness. Less consistent is the diameter of the sub-pores at the base of each hemispherical void which vary both in regularity of shape and size, ranging from <50 nm to 250 nm across the sample due to the aforementioned irregular contact area between microspheres and the PVDF surface, Figure 6.

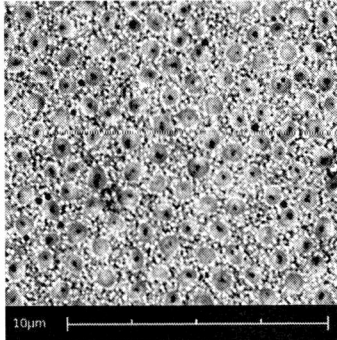

Figure 6 Nanoporous Ag deposit on TiPE O502 dip-coated PVDF, t = 60 min

When the deposition time is further extended to 120 min, a thicker Ag layer is produced through continued autocatalytic deposition, as would be expected. An average hemispherical void mouth diameter of 950 nm can be measured, Figure 7, equating to a metal layer thickness of approximately 344 nm. As metal growth continues upwards, the flat, essentially non-templated, regions of metal between pores decrease in size and the overall metal surface area is increased, providing potential advantages for a range of intended applications such as supercapacitors (13), fuel cells (14) and surface enhanced resonance Raman techniques (15). For a monolayer of pores / voids, a deposition period capable of producing a metal layer of thickness equal to one half of the microsphere diameter is desirable for optimal surface area. As the Ag deposit produced in this instance is less than the 500 nm radius of the microspheres used in the template, it is clear that deposition times greater than 120 min are required to produce a metal layer with optimal surface area.

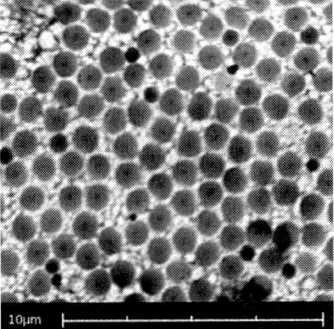

Figure 7 Nanoporous Ag deposit on TiPE O502 dip-coated PVDF, t = 120 min

Conclusion

We have reported the photocatalytically initiated electroless deposition of metals with highly ordered nanometer scale porosity onto TiO_2 sensitised PVDF membranes. The use of a PVDF substrate presents a number of process problems such as the lack of facility to anneal the TiO_2 coating, disruption of the microsphere self-assembly process and reduced photocatalytic nucleation. Despite these difficulties, we demonstrate the formation of porous Ag metal films on membrane substrates, with thickness determined by the total deposition period. The initial nucleation and growth patterns of the produced metal films are seen to be similar to those previously described for the same, templated PIED process on glass substrates (4, 7).

Potential applications of these membranes, such as analytical and process separations and electrode / solid electrolyte composites for fuel cells, require that the porosity induced in the deposited metal must penetrate through to the PVDF membrane. This is achieved as no metallisation occurs at the points of contact between microspheres and the membrane surface. Metallised separation membranes may have potentially wide applications in analytical separation technology with work in our laboratories focusing on their application in electrochemically controllable and enhanced actinide separations in new nuclear reprocessing platforms. This will be the subject of the next paper in this series.

Acknowledgments

We thank the Royal Society of Chemistry UK, EPSRC (Award No EP/I002928/1) and The Lloyds Register Educational Trust (LRET) for financial support. The LRET is an independent charity working to achieve advances in transportation, science, engineering and technology education, training and research worldwide for the benefit of all.

References

1. M. A. Bromley, C. Boxall, S. Galea, P. S. Goodall, S. Woodbury, *J. Photochem. Photobiol., A*, **216**, 228, (2010).
2. M. A. Bromley, C. Boxall, *UK Pat. Appl. WO/2012/042203*, (2012).
3. M. A. Bromley, C. Boxall, in *Adv. Chem. Res.*, Vol. 13 (Ed: J. C. Taylor), Nova Publishers, (2011)
4. M. A. Bromley, C. Boxall, *Electrochem. Commun.*, **23**, 87, (2012).
5. A. Mills, S. Le Hunte, *J. Photochem. Photobiol., A*, **108**, 1, (1997).
6. P. N. Bartlett, P. R. Birkin, M. A. Ghanem, *Chem. Commun.*, 1671, (2000).
7. M. A. Bromley, C. Boxall, *J. Mater. Chem. A*, **under review**, (2013).
8. G.-J. Yang, C.-J. Li, F. Han, X.-C. Huang, *J. Vac. Sci. Technol., B*, **22**, (2004).
9. J. C. Yu, J. Yu, W. Ho, J. Zhao, *J. Photochem. Photobiol., A*, **148**, 331, (2001).
10. Z. Wang, U. Helmersson, P.-O. Käll, *Thin Solid Films*, **405**, 50, (2002).
11. M. A. Bromley, C. Boxall, *MRS Proc.*, **1409**, (2012).
12. K. R. Denison, C. Boxall, *Langmuir*, **23**, 4358, (2007).
13. G. S. Attard, J. M. Elliott, P. N. Bartlett, A. Whitehead, J. R. Owen, *Macromol. Symp.*, **156**, 179, (2000); V. D. Patake, S. S. Joshi, C. D. Lokhande, O.-S. Joo, *Mater. Chem. Phys.*, **114**, 6, (2009).
14. W. Yuan, Y. Tang, X. Yang, Z. Wan, *Appl. Energy*, **94**, 309, (2012).
15. W. E. Smith, C. Rodger, in *Encyclopedia of Spectroscopy and Spectrometry (Second Edition)*, (Ed: L. Editor-in-Chief: John), 2822, Academic Press, Oxford (1999)

Author Index

Acevedo-Peña, P.	81	Khunsin, W.	53
Armstrong, E.	53	Kim, Y. G.	113
Atsushiba, H.	99	Ku, J.	45
Aureau, D.	93, 105		
		Lee, K.	45
Boxall, C.	123	Lotty, O.	25
Bromley, M. A.	123	Lynch, R. P.	65
Buckley, D. N.	65		
		Matsuda, H.	99
Chazalviel, J. N.	3	McSweeney, W.	25
		Mercier, D.	93
Dornhege, M.	65		
		Njel, C.	93
Enomoto, M.	99		
Etcheberry, A.	93, 105	O'Dwyer, C.	25, 53, 65
		Okayama, C.	99
Fukumuro, N.	99	Osiak, M.	25, 53
Geaney, H.	25	Park, S.	45
Gebregziabiher, D.	113		
Glynn, C.	25	Quill, N.	65
Gonçalves, A. M.	93	Quiroga-González, E.	25
González, I.	81		
		Rotermund, H. H.	65
Hasegawa, A.	99	Ryan, K. M.	53
Hervier, A.	105		
Holmes, J. D.	25	Sakamoto, S.	99
Hong, K.	45	Sotomayor Torres, C. M.	53
		Stickney, J. L.	113
Jones, K.	25		
		Tsang, C. F.	113
Kang, S.	45		
Kennedy, T.	53	Yae, S.	99